Managing New Product Innovation

Managing New Product Innovation

Proceedings of the Conference of
the Design Research Society
Quantum Leap:
Managing New Product Innovation
University of Central England
8–10 September 1998

BOB JERRARD
University of Central England

MYFANWY TRUEMAN
University of Bradford

ROGER NEWPORT
University of Central England

UK   Taylor & Francis Ltd, 1 Gunpowder Square, London, EC4A 3DE

USA   Taylor & Francis Inc., 325 Chestnut Street, Philadelphia PA 19106

British Library Cataloguing in Publication Data

A catalogue record for this book is available from the British Library.
ISBN 0-7484-0859-2 (cased)

Library of Congress Cataloguing-in-Publication Data are available

Cover design by Hybert Design and Type
Printed and bound by T.J. International Ltd, Padstow, UK
Cover printed by Flexiprint, Lancing, West Sussex

Contents

Introduction

Bob Jerrard, Myfanwy Trueman and Roger Newport

This book presents a selection of papers from the Design Research Society conference **Quantum Leap**: *managing new product innovation*, held at the University of Central England from 8th to 10th September 1998. The event was held in collaboration with the University of Bradford Management Centre, the Design Council, the Department of Trade and Industry, and the Chartered Society of Designers.

THE MEANING AND VALUE OF DESIGN

It has been said that "a nation's competitiveness depends on the capacity of its industry to innovate" (Porter, 1990) but it is increasingly difficult for industrial companies to produce successful new products in the face of intense global competition and today's turbulent business environment. Hard pressed managers have to make difficult decisions in a complex world of perpetual change and a climate that has been described as "permanent white water" (Bartlett and Ghoshal, 1995). Can design be used to support these decisions and sustain a competitive advantage?

To be successful, innovative products must take into account opportunities provided by new technology and materials on the one hand but they must not loose sight of the customer on the other. According to leading designers Seymour Powell (1993), there is a need to take "no more than half a *quantum leap* at a time" to stay in touch with customers and others working on each new project. If designers loose touch with the values, beliefs and needs of the market place it may be difficult to get new products accepted, and if companies do not have an integrated project team, problems are likely to occur even before product launch. This might be related to the 'millennium dome effect', people could not fail to be impressed by the design and technological achievement of London's millennium dome but there was a great deal of uncertainty in July 1998 about its value and purpose. To counter this, Fujimoto (1990) points out that successful products require 'integrity' — a blend of *productability*, *usability* and *appropriateness*.

INNOVATION MANAGEMENT

An organisation that is well managed has a commitment to planning. Such plans will involve strategies, a structure and a mission as well as market or product preference. Innovative and ambitious organisations must show a commitment to a regular accumulation of ideas, and a policy that is supportive of new projects but is balanced by the management and control of risk. *Quantum leap* in this context describes a major breakthrough or sudden advance, which may achieve a competitive advantage on anything from five minutes ahead of the nearest rival to the acquisition of a completely new market.

This balance of innovation and control may only be achieved by reconciling the different factions in a company, its business and marketing environments to form strategic alliances. Instigating a design focus may facilitate such a process but in order to do this we need to develop a greater understanding of the meaning and potential of design.

New product innovation is considered mostly in terms of technological innovation, but whereas new technology does not necessarily have to be part of design innovation, most new technologies cannot be implemented without it. Innovative design however, still seems to contain more market risk than innovative technology.

Risk is mostly associated with management decisions about new designs or new technologies made without appropriate information on continuous contextual change. A range of techniques from the designer's intuition to speculative market testing provides shortcuts to its reduction. Design research and new information technologies can provide metaplanning information towards continuous contextual fit, and reduce currently acceptable levels of risk.

CONFERENCE AIMS

This conference aims to bridge the gaps between some disparate factions, and:
- To encourage new initiatives and joint ventures between the research, design and business communities.
- To investigate the management of product innovation to promote economic competitiveness.
- To provide a forum for discussion and dissemination of research findings and industrial practice.

QUANTUM LEAPS

The theme examines: how a *quantum leap* or breakthrough can occur in products, management systems or company culture, by using the many facets of design; and ways in which design can influence the shared process of innovation and decision making. It will also make reference to the ways in which we manage and measure product success from getting established in a market and recouping investment, to increases in market share, identity, image and product values. These financial and operational measures allow for a systematic process to feed strategy. Rational decision-making occurs at strategic levels where choices may be influenced by using design to communicate, interpret and integrate ideas thereby unifying an apparent multiplicity of goals and types of people within an organisation. Rationality might also be used to overcome uncertainty in the market place if used in collaboration with marketing research.

If design does become central to strategic decision-making in companies, this can in itself be perceived as a threat to traditional work practice and bureaucratic control, particularly where there is a direct link between designers and the customer. However the mobilisation of design ideas can feed corporate commitment to design so that the vision of the designer or project team becomes a change agent for the whole company. These ideas form a central tenet of the conference proceedings since there is a clear need for collaborative relationships to be developed between the design, research and business communities, each of which are likely to work to rather different agendas. Consequently some of the main issues raised by this Design Research Society conference include:

- The kind of design techniques that are used in organisations to produce successful, innovative new products.
- How companies with bureaucratic structures can benefit from the uncertainty presented by innovative design ideas.
- How company managers can take advantage of design research and how this works in practice.
- How research into design activities and the product development process affect organisational change.
- How successful companies use design in order to reduce risk and uncertainty in developing innovative new products.
- How design can be used creatively to make a major breakthrough or a *quantum leap* in successful innovation practice.

To this end these proceedings include a range of product and company case studies, surveys, research papers and collaborative ventures. Contributors to this book are drawn from design practitioners and research teams at universities and business schools who are working to discover how new ideas in design can support future industrial strategy.

ABOUT THE EDITORS

Dr Bob Jerrard is Professor of Design Studies at UCE's Birmingham Institute of Art and Design. His research interests include aspects of management and psychology and he has published internationally on design theory and technology diffusion. He also teaches and directs research in these subjects. Recent publications (1998) include: Quantifying the Unquantifiable: An Enquiry into the Design Process, *Design Issues*, Vol.14, Spring, pp. 40-53. (Cambridge, Mass.: M.I.T. Press).

Dr Myfanwy Trueman is the Heinz Lecturer in Marketing at Bradford University Management Centre. Her *new topology*, based on doctoral research at Bradford University (Management), uses design attributes to determine the value, image, process and production (VIPPs) of new products. Recent publications (1998) include; Long Range Planning, In *European Journal of Innovation Management*, Vol.1, No.1, pp. 44-56.

Roger Newport is Professor of Industrial Design and heads the School of Design Research at UCE's Birmingham Institute of Art and Design. His background is in product design practice, overseas development in design, and teaching design theory, history and practice. Recent publications (1998) include: *Connecting Element,* UK Patent GB 2292204B, United Kingdom Patent Office.

ACKNOWLEDGEMENTS

The Editors would like to thank the Design Research Society, the University of Central England in Birmingham, the University of Bradford Management Centre, the Design Council, the Department of Trade and Industry Innovation Unit and the Chartered Society of Designers for their support, together with the contributors to this book. We would particularly like to thank Denise Carpenter and Chris Dyke for their patient preparation of the final document.

REFERENCES

Bartlett, C. A. and Ghoshal, S., 1995, Rebuilding Behavioural Context: Turn Process Re-engineering into People Requirements, In *Sloan Management Review*, Fall. (Cambridge Mass.: Alfred P. Sloan School of Management, M.I.T).

Fujimoto, T., 1990, *Growth of International Competition and the Importance of Effective Product Development Management and the Role of Design.* In Proceedings of Product Strategies for the 1990's Conference, (London: the Financial Times).

Porter, M. E., 1990, The Competitive Advantage of Nations, In *Harvard Business Review*, Vol.90, No.2, (Harvard, Mass.: Graduate School of Business Administration, Harvard University).

Seymour, R. and Powell, R., 1993, *DTI Enterprise Initiative: Design to Win.* (Department of Trade and Industry, UK).

System-Operated Product Development and the Craft of Integration

K. L. Bull

System-Operated Products (SOP's) such as personal digital assistants and portable digital communication devices, are single elements within larger interdependent systems. They rely upon the exchange of electronic information such as sound, text, images and video. These systems are becoming a significant feature of post-industrial societies but may ultimately challenge the traditional relationships people have with time and place. This is a result of the increased demand for near instantaneous information exchange that is not subject to geographical constraints. The problem is that this often requires the use of complex computing and telecommunication technologies. These are frequently in the form of 'hidden layers' of technology (e.g., cellular transmission technologies and service providers) which may make it harder for the user to develop a suitable understanding of a product's optimal function and operation.

The discussion outlines some potential opportunities which may help to emphasise the qualities characteristic of the craft-based approach to product development, and suggests how this may be achieved through the wider utilisation of the industrial designer during System-Operated Product Development (SOPD).

The objective is to identify how the designer may assist collaboration amongst specialists to ensure that the user becomes a central participator within SOPD. It is suggested that this may be partly achieved by the use of flexible information and manufacturing technologies. The need for an adaptable framework for SOPD is also considered, which promotes the appreciation of whole systems and their products rather than the isolated development of products. The paper examines how these suggestions might be strengthened by allowing the industrial designer to act as a non-specialist across disciplines in order to apply, from an objective position, insight and intuitive skills to areas of product development which are primarily rational and scientific. It also reinforces the idea that the industrial designer could play a significant role as an integral and non-hierarchical member of an SOPD team.

INDUSTRIAL DESIGN AND TECHNOLOGICAL CONVERGENCE

The implementation of industrial design methods is subject to constant change. It is therefore important to consider the combined theory and practice of design as an evolutionary process that accommodates new technologies and changing social values. This was first emphasised by the Design Methods Movement (DMM) during the 1960s. One of its founder members, John Chris Jones (1985), identified the need to appreciate the product by understanding its whole. He talked primarily about the social, economic and political basis of the existence of a single product in order to address human needs (Fig. 1). This need for change was recognised as being a result of the increased complexity of new products brought about by technological developments. This is significant today because of the convergence of computers, telecommunications and electronic information systems. This may eventually alter our perception of the 'product', which would be likely to incorporate many hidden layers of technology and function; and drive opportunity for new types of electronic information-based products and services (Fig. 2). It can be argued that this will broaden the parameters of industrial design to encompass a wider range of specialist skills and address implications which reach far beyond the self-contained product.

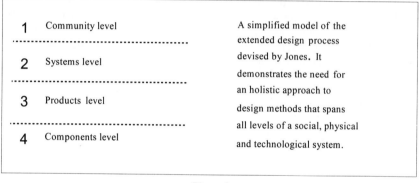

Figure 1

   Although aware of these issues in the 1960s, the DMM was unable to optimise the practice of industrial design. This was partly due to their attempt to place greater emphasis upon the use of logical and scientific methods over the intuitive and irrational skills fundamental to design. This demonstrates the difficulty of combining the theory and practice of design methods. This paper aims to highlight that dilemma in the context of the new design complexities associated with the 'information age' and to reinforce the idea that many of the motives of the DMM are extremely significant today, within the context of industrial design and system-operated product development (SOPD).

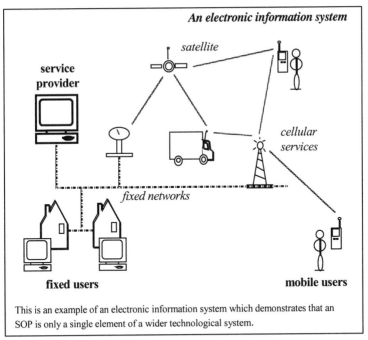

An electronic information system

This is an example of an electronic information system which demonstrates that an SOP is only a single element of a wider technological system.

Figure 2

THE EVOLVING ROLE OF DESIGN METHODS

During the preindustrial era the user was often the main indicator of needs during the design process. Products were often well fitted to their context of use. Mass-manufacturing techniques have made this advantage harder to maintain. The need to push large product volumes and new technologies into general markets has resulted in the design process becoming mainly focussed upon product opportunity rather than the user. This has resulted in the user's reconciliation with mass-manufactured goods. It is however, possible to align mass-manufacture with the modernist ideal that standardisation would ultimately lead to an well-ordered society. Walter Gropius believed that standardisation would not restrict the individual but allow them to develop more freely. It can now be seen that this was at the expense of humanistic values (Zaccai, 1995).

A consequence of mass-production is that it suppresses individuality. This is supported by Andrea Branzi who suggests that our current technological climate now has the potential to promote individuality and this is fostering the search for flexibility (Thomas Mitchel, 1993). This is a result of new technological opportunities such as mass-customisation and the use of integrated technologies. These are characteristic of the 'information age' which has according to the social scientist Daniel Bell, evolved from industrial society as a consequence of information technology and the centrality of 'theoretical knowledge' (Bell, 1974).

It challenges many existing social, economic, and political frameworks. Transnational Corporations for example can reach beyond the government, policies and economic restrictions of their country of origin (Lyon, 1994). This view is supported by Thomas Mitchell (1993), a design theorist, who suggests that "The transition to the post-industrial has necessitated a change in the focus of the design process from physical form to users experience, from mass production to custom adaption, and to paraphrase Morris – from having machines, and industrial technologies generally, be our masters to having them be our servants".

It is likely that a change in emphasis is needed to adapt to the needs of the 'information society' which constitute the social, political and economic factors that are shifting from a reliance upon mass-production and standardisation, to individuality and interconnectivity enabled by the flexibility of communications and information exchange. This may alter people's relation with time and space and ultimately their social values. Therefore, it is probable that industrial design and SOPD could benefit from a more timely review of the motives of the DMM of the 1960s.

THE DESIGN METHODS MOVEMENT

During the 1960s it was identified that designers could no longer solely rely on their ability to focus upon the product as the centre of the design task. Due to technological developments and the implications of mass-production, interest had to be shifted from the design of hardware and form to the consideration of human needs. This required a new look at the subject of design methods.

Design methodology is defined as "the study of the principles, practices and procedures of design in a rather broad and general sense. Its central concern is how designing both is and might be conducted. This concern therefore includes the study of how designers work and think; the establishment of appropriate structures for the design process; the development and application of new design methods, techniques, and procedures; and the reflection on the nature and extent to design knowledge and its application to design problems" (Cross, 1985).

The DMM was founded in recognition of this need and the new types of problem occurring in design at that time. Jones identified that the traditional design process could no longer support the increasing complexity of problems. These were a result of the 'artificial', such as traffic congestion and the shortage of public services. He suggested that the need for new methods was stimulated by the "failure to design for conditions brought about by the products of designing" (Jones, 1992). He indicated "that the design process should be extended from its concern with products, to include the design of systems" (Cross, 1992). This view was also supported by the industrial designer Bruce Archer (1985). He suggested that the emphasis within post-industrial culture should be to consider the design of the whole system of which a proposed product is a part, rather than considering the product as a self-contained object. He also identified the importance of involving a wide range of specialists (such as those involved in ergonomics, cybernetics and management sciences) during the development process.

Thomas Mitchell, (1993) outlines the aims of the DMM and recognises the need for improved collaboration within the design process. "Specifically design methods were developed to permit many people to collaborate in the design process, in place of the reliance on a single person's ability to know and effectively synthesise all of the information relevant to a design task". He highlights the principal characteristic of the early years of the movement indicating that design methods were aimed towards incorporating systematic forms of information into the process where traditionally it had been mostly concerned with the intuitive techniques adopted by individual designers.

Christopher Alexander, a key member of the movement, argued that its objectives could not be reached through an 'unselfconscious' craft-based approach characteristic of traditional design activity. He believed that this should give way to the 'self-conscious' professionalisation of design (Lawson, 1990). Jones also refers to this, he defines the limitations and rigidity of design process during the industrial era as opposed to the traditions of craftsmanship, where the designer actually makes and evolves the product through experimenting with form. He describes 'design-by-drawing' as a process typical of the industrial era, where the designer is not responsible for the actual making of the product. Whilst drawing is a useful mechanism to manipulate ideas, such as form, and to visually comprehend alterations, its greatest disadvantage during the design process is that non-aesthetic or non-visual problems (such as social implications of estate planning) are not exposed (*ibid.*). Jones and Alexander recognised this and sought means to identify and resolve non-visual problems through different models of representation.

Jones believed that the use of traditional drawing techniques within the design process prevented innovation at the systems level. He proposed the need for 'multi-professional' designers who made intuitive decisions based upon prior knowledge and experience (Jones, 1992) and emphasised the need for finding new methods that reach across the physical and social levels of designing (*ibid.*) He suggested that "The kinds of design skill which are called for in using the newer design methods, ... are suited to collaboration, the sharing of responsibilities between users and experts, and to designing imaginatively in a collective process, as was the case in craft evolution". He also recognised the need for the design process to be more open to judgement and for not elevating the designer to a position of deity (Thomas Mitchell, 1993).

THE DESIGN PROCESS

The systematic problem solving techniques adopted by the DMM were taken from management and computer sciences. The aim was to use them to assess problems and aid in the development of design solutions. This encouraged the development of models of the design process. The principle was based upon dividing the process into two key areas, 'problem definition and problem solution' (Buchanan, 1995). This corresponds with the linear model characterised by the 'analysis, synthesis and evaluation' phases of design (Hickling, 1982). Allen Hickling suggested in *'Changing Design'* that these models were difficult to use as a result of oversimplification.

Jones defined the systematic approach as primarily a way of keeping logic and imagination separate by external means, leaving the mind free to develop ideas without the limitations and confusion of working through the analysis. The analytical information would be stored outside the mind and separate from the solution until they could be logically related (Jones, 1985). He identified that within early systematic methods it was not easy to move between the rational, systematic methods and the creative and imaginative aspects of the design process. Eventually Jones changed his entire philosophy on design methods and thought that the whole process had become over-rationalised, eliminating the fundamental characteristics of intuition and imagination (Thomas Mitchell, 1993).

Many members of the DMM advocated the need to work in a systematic way but they did not deny the qualities of the craft-based approach. Nigel Cross highlights these qualities saying that "the preindustrial craftsmen seem to have been capable of producing objects whose forms (or 'designs') tend to exhibit a remarkable subtlety within an apparent simplicity. This simplicity often masks the complexity of the relationship between form and function that the object embodies" (Cross, 1992).

Horst Rittel (Rittel and Webber, 1985) sought an alternative framework to the practice of design. It had been hoped that the linear method would provide a "'logical' understanding of the design process". Rittel identified the notion of 'Wicked problems'. He argued that design problems are naturally ill-defined and indeterminate. 'Tame problems' he defines as well structured problems, where all the necessary information for solving them is available. These fitted well into the systematic approach to designing. Rittel elaborated that 'wicked' problems required the designer to explore the relevant information about the problem through an argumentative process based upon previous experience and participation with other specialists. The objective was to allow a solution to emerge gradually.

DESIGNER AS INTEGRATOR

This analysis illustrates the need for the designer to act as a collaborator and integrator during the development process. The architect Peter Behrens, for example, was employed as a non-technical specialist within the manufacturing process by the Allgemaine Elektrizitats-Gesellschaft (AEG) company in 1907. His responsibility was to improve the visual appearance of mass-produced electrical products. He was unable to rely on previous knowledge to work with a full understanding of technologies and skills relating to the development of mass-produced electrical goods. Behrens and other industrial designers at that time, concentrated efforts upon the development of a 'machine aesthetic' which was intended to overcome some of the shortcomings of mass-production. The product development process was typically left to technical and marketing specialists and this division eventually resulted in the design of superficial products. Poor design quality emerged because the team could not appreciate the balance of "combined human senses" (Zaccai, 1995). Zaccai suggested that the designer should have responsibility for all aspects of a product's development to gain a clearer understanding of the natural balance point amongst human needs. He also indicates that the designer may not need to be very knowledgeable in all the specialisms of product development.

This demonstrates the need for an integrated approach to design and reflects the DMM objective for better collaboration within the design process. Whilst the designer should not assume hierarchical importance within the process, their contribution may be usefully integrated amongst disciplines. Buchanan indicates that this is still an unresolved issue. He explains that specialised studies in the arts and sciences have become fragmented and suggests that the commonalties between the disciplines are being lost. Buchanan argues that without the combination of "understanding, communication, and action" it is difficult to implement knowledge and generate ideas towards the enrichment of human life (Buchanan, 1995).

THE FUTURE OF A CRAFT BASED APPROACH

Thomas Mitchell (1993), however, proposes that we are already approaching a state in the maturity of technology where products can be produced with the character of handicrafts – "high quality, well adapted to their context of use and even customised and individualised". It is evident that there are a number of techniques which support this ideal and which are closely aligned to the motives of the DMM. They include methods such as participatory design, collaboration, user-centred system design, humanware, and transparency.

'Humanware' has been adopted by some leading Japanese electronics manufacturers. It involves the organisation shifting marketing strategies towards a "needs orientated approach which aims to make technology more intelligent, more flexible for users of different cultural backgrounds and generally to consider the social context of their products" (Pilditch, 1989). It shares some common goals with User-Centred System Design, pioneered by Donald Norman and Stephen Draper. They emphasise the need for using cognitive engineering to improve decision making and the choice of the correct conceptual models for a products use. This requires the identification of goals and psychological intentions in relation to user needs. This is not dissimilar in approach to 'transparency,' a method described by Thomas Mitchell (1993) which explains how the "intangible processes of use guide the development of tangible artefacts". Within transparency the social context and methods of use are a primary consideration before the development of a technological specification.

The Human-Centred Movement (Gill, 1996) is specifically concerned with the design of technology. It has two major objectives; 'human-machine symbiosis' and 'socially useful technology'. One motive is to "aim towards placing the human needs, purpose, skill and creativity at the centre of technological systems". They appreciate the importance of collaboration within their activities.

The need for 'participation' within the design process is also recognised Sonnenwald (1996) supports the need to explore the design context and expands upon the diversity of specialists who may participate within a design project. In response to this Sonnenwald suggests the need to expand design methods to encompass communication roles and strategies and describes how this might be achieved through a prescriptive framework which utilises computerised technology to support the integration and formulation of knowledge during the design process.

In support of 'participation' Reich, Konda, Monarch, Levy and Subrahmanian discuss the importance of conducting an "objective evaluation of need as integral to the design process". They propose that traditional social science techniques may not be the most productive way of gaining information about needs and stress the importance of providing the user with an active role during the design process (Reich *et al.,* 1996).

This has been practised by Design Continuum Inc. in the United States. It has been working on a programme called 'SEED' – 'Science, Ergonomics, and Early Design.' The group focuses on the opportunities for new products and services. They have found that ideas usually appear at the intersection of technology, social trends and human factors, and try to identify unmet user needs and synergistic technologies. They believe that while market research is useful it is often a poor predictor of future needs and opportunities. They consider it problematic to determine design characteristics by asking consumers to quantify their needs and stress that intended consumers should experience the possible benefits. The approach includes using techniques such as computer simulations and rapid prototyping (Zaccai and Badler, 1996).

These examples illustrate how design methods might support the qualities associated with craft based design within advanced technology product development to deliver outcomes which are closely aligned to user needs.

CONCLUSION

A principal advantage of craft-based design was identified as 'user participation'. The opportunities created by new manufacturing techniques during the industrial era, however, contributed to the separation of the designer from the process of making and promoted the departure from individually crafted products to mass-manufactured goods. This paper proposes that the current technological climate might now support craft-based ideals through information technologies and automated manufacturing techniques. These can support user participation at all levels of the design process and contribute to artefacts that have a closer affinity to their user and social contexts.

The DMM highlighted the importance of meeting user needs by appreciating the whole system because of the nature of new design problems at that time. The growth of a service based culture, identified as adding to the complexity and sophistication of many existing and new products, makes an holistic approach especially relevant when dealing with information based products which have technological systems that contain complex interdependent components. For example, a car navigation system may require near-synchronous access to an external database of information whilst the vehicle is moving. This would be strengthened by specialist collaboration and participation during the design process. In this situation the designer may usefully act as a non-hierarchical integrator within SOPD to aid communication amongst specialists and help match user needs with technological potential and environments of use. For example, a personal communication device may be operated across varying local, regional or even international boundaries and environmental constraints, such as censorship, noise, or cultural differences. It was recognised that a designer will often not have experienced knowledge in all development specialisms but may utilise their skills of communication, prior experience, intuition and insight to help produce solutions to the ill-formulated or 'wicked' problems which are characteristic across the design spectrum.

The following summary proposes areas for further investigation in the field of SOPD activity. The aim is to identify a set of methods which are closely aligned to some of the qualities associated with craft-based design.

- The industrial designer could usefully contribute as an *integral* member of the design team, aiding communication and the resolution of 'wicked' problems amongst disciplines within SOPD;
- SOPD should be a *collaborative* activity which involves user participation;
- An *holistic* approach to SOPD may encompass the broader implications of products, users and their associated systems;
- Appreciation of the *social context* of new products is valuable but could be imperative to the design of SOP's which have complex features and interdependent functions;
- The industrial design specialist might also act as a *non-specialist* across disciplines and be in a good position to make an objective contribution to the SOPD team;
- SOPD could benefit from a *flexible and adaptable approach* (to accommodate social and technological change) which combines both systematic and intuitive design methods;
- The development of appropriate *system and conceptual models* for use might enhance SOPD;
- *Visual models of representation* may usefully identify and support the search for solutions to hidden or intangible problems associated with SOPD.

REFERENCES

Archer, B., 1985, Systematic Methods in Design. In Cross, N., *Developments in Design Methodology*, (Chichester: John Wiley and Sons).

Bell, D., 1974, *The Coming of Post-Industrial Society: a venture in social forecasting*, (London: Heinemann Educational).

Buchanan, R., 1995, Wicked Problems in Design Thinking in Buchanan, R., and Margolin,V., *The Idea of Design: A Design Issues Reader*, (Cambridge USA: The MIT Press).

Cross, N., 1985, *Developments in Design Methodology*, (Chichester: John Wiley and Sons).

Cross, N., 1992, The Changing Design Process. In Roy, R., and Wields, D., *Product Design and Technological Innovation*, (Milton Keynes: Open University Press).

Gill, K., 1996, The Human-Centred Movement: The British Context. In *AI & Society*, (London: Springer International).

Hickling, A., 1982, Beyond a linear iterative process? In Evans, B., Powell, J., Talbot, R., *Changing Design*, (Chichester: John Wiley and Sons).

Jones, J. C., 1985, A Method of Systematic Design. In Cross, N. *Developments in Design Methodology*, (Chichester: John Wiley and Sons).

Jones, J. C., 1992, *Design Methods*, (2nd Edn.), (New York: Van Nostrand Reinhold).

Lawson, B., 1990, *How Designers Think: The Design Process Demystified*, (2nd Edn.), (London: Butterworth).

Lyon, D., 1994, *The Information Society: Issues and Illusions*, (Oxford: Polity Press).

Pilditch, J., 1989, *Winning Ways*, (2nd Edn.) (London: Mercury).

Reich,Y., Konda, S., Monarch I., Levy, S., Subrahmanian, E., 1996, Varieties and Issues of Participation and Design, In *Design Studies*, Vol.17, No.2, (London: Elsevier Science), pp.165-180.

Rittel, H. and Webber, M., 1985, Planning Problems are Wicked Problems. In Cross, N., *Developments in Design Methodology*, (Chichester: John Wiley and Sons).

Sonnenwald, D., 1996, Communication roles that support collaboration. In *Design Studies*, Vol. **17**, No.3, (London: Elsevier Science), pp. 277-301.

Thomas Mitchell, C., 1993, *Redefining Designing: From Form to Experience,* (New York: Van Nostrand Reinhold).

Zaccai, G., 1995, Art and Technology In Buchanan, R. and Margolin,V., *Discovering Design,* (London: Chicago Press).

Zaccai, G., Badler G., 1996, Insights on the practice of consulting: New Directions for Design. In *Design Management Journal*, Spring, (Boston), p. 57.

Heroes and Villains? The Contradictory and Diverse Nature of Design Management

Rachel Cooper and Mike Press

This paper considers the rise of design management within education and practice, locating it within the political economy of the 1980s. After considering its contradictory nature as a management tool which aims to constrain the benefits that design offers, the paper then explores its diverse nature in practice. Based on four detailed case studies, a critique emerges which questions the professional coherence of design management. The authors conclude that a greater sense of definition and identity be applied to both design management and design education, and propose more specialised definitions and courses of study. They argue that 'design management' is a term which may have outlived its usefulness.

THE RISE OF DESIGN MANAGEMENT

The 1980s gave us a number of fashionable ideas claiming to get Britain back on its feet, get innovation straight in our heads, and get more and better products into our eager hands. These ideas include Thatcherite free-market economics, business process re-engineering, enterprise culture and design management. All were radical, ambitious, interdependent and in their various ways, all were failures. This failure stemmed from their status as articles of faith, as ideologies, which were all born of a particular political moment. That moment has now passed, and it is therefore necessary to analyse the reasons for failure and consider alternatives.

Design management spiralled up on the thermals generated by the decade of conspicuous consumption, seemingly connecting Britain's landscape of industrial blight with the blue sky thinking of our much-lauded designers. The free market orthodoxy of the 1980s claimed to offer the only path to an innovative economy. The role of government, which in the past had concerned itself with regulation, planning and targeting central funds at R&D was now, simply, to get out of the way, offering less regulation, less or indeed no planning, and far less state investment. In place of these was encouragement to foreign investors, to small business and to cost-cutting. A key part of this strategy was the encouragement to use design, through grants and through the exemplars and evangelising of the Design Council. It was the 1980s that saw the emergence of design management: research and prescriptive measures to encourage firms to make use of design as a competitive weapon.

The term itself actually dates from the 1960s with the establishment of the RSA (Royal Society of Arts) Design Management Award in 1965 and the publication of Michael Farr's book 'Design Management' the following year. These developments reflected efforts within the consultancy sector to raise the professional status of design by offering services which addressed a wider range of product development and corporate strategy issues. An increasing concern with systematic design methods, as developed by Bruce Archer and John Chris Jones, was linked to efforts by the profession to distance themselves from the 'romantic hero' model of the designer.

But it was in the 1980s that design management became a political project. The Society of Industrial Artists and Designers' (SIAD) Design Management Group was formed in 1981, and the London Business School championed the subject academically through its MBA programme and a series of seminars published by the Design Council (Gorb, 1988 and 1990). Design management was soon embraced by design education and courses were established within design faculties at DeMontfort, Middlesex, Staffordshire and other universities, together with the Royal College of Art. Design management symbolised the hegemony of business management in the new political economy of education. It was possible to sweep aside the politically incorrect histories of design with their emphasis on William Morris and the Bauhaus with their leftist leanings and assert that design had really been a form of management all along. As Wendy Powell, formerly of DeMontfort University, argued: "if you really look at the best design thinking and the best management thinking they have a lot in common" (Powell, 1995, p.45). There was of course another driving force behind the new design management courses: they were cheaper to deliver.

As a means of legitimising design in terms of the new ascendant free-market ideology, design management had a number of key objectives:

- Through research it sought to demonstrate the commercial value of investment in design to industry and commerce (Service *et al.*, 1989; Roy and Potter, 1993). The recently restructured Design Council took on the commissioning of such research as its main mission.
- Management practices and strategies were identified in exemplar companies from which approaches to policy development in corporate design management were elaborated (Oakley, 1990; Cooper and Press, 1995).
- New means of discussing and disseminating the design management message were developed, including the Design Management Journal in 1989, Co-design Journal in 1994, the European Academy of Design in 1995.
- Through new course development at undergraduate and postgraduate levels to educate students in the skills, knowledge and understanding relevant to design management.

The authors of this paper have been among those who have been commissioned to do the research and write the design management textbooks. We have sat on editorial boards, run and attended the conferences, set up, managed and examined a few of the courses in our time. From our active role in the field arises a view that design management is in need of far more critical reflection on its purpose and direction. Indeed, there is a case for scrapping the term entirely as an unhelpful and politically defined catch-all for a number of very distinctive and separate practices and issues. In design education there is a vital need to empower students to be proactive in business and culture, but design management, as it has defined itself in a number of institutions, does not constitute a body of knowledge or set of critical aspirations that makes this possible.

Our critique focuses on two themes. First, the contradictions which lie at the heart of design management as a concept which combine a diminished sense of purpose and ambition on the part of designers with the imperatives of managerial control driven by short-term interests to produce a system which actually reduces the influence of creative thinking and action. Second, the variety of 'design managements' that are revealed when you examine what design managers actually do, suggesting the need for discrete areas of research, education, training and professional recognition. As Nick Hornby observed in his novel 'High Fidelity', Bill Wyman and Keith Richards both play the guitar, but you wouldn't say that they do the same job.

THE CONTRADICTIONS OF DESIGN MANAGEMENT

According to Rodney Fitch "it is probably true that the (Thatcher) government banged the drum for design-led innovation like no other for 100 years". As a driving force behind the new design management we first have to establish whether all the drum banging through seminars, videos, DTI and Design Council literature did actually wake up industry to the benefits of design.

Research suggests that few firms took any notice of it. According to a 1996 survey of thirty five manufacturing firms by Tom Fisher (1996) at Sheffield Hallam University, only three had made any use of DTI literature in terms of developing procedures to manage New Product Development (NPD). However, this research does indicate that certain elements of 'good' management practice in NPD and design management are fairly widespread. The use of cross-functional teams and the adoption of set procedures in NPD can be seen in the majority of companies surveyed. A further part of the survey was to ask designers who worked in such teams whether it enhanced or constrained their creativity. In the majority of cases they considered that it increased their ability to be creative. At first sight then, British companies appear to be marching to the drumbeat of the design management message, wherever they're hearing it from, and designers are unlocking more of their creative talent. But let us not be too hasty in getting out the bunting and singing a few choruses of 'design management's coming home'.

According to Peter Wickens (1995) "more nonsense is written about teamworking than almost any other concept". In his view cross-functional project teams only work if two other things are in place. The first is a commitment to permanent teamworking that infuses the entire company, and the second is a real commitment to job security. Given the continued tenacity to hierarchies in most organisations, and the continued fashion for corporate downsizing and short-term contracts, the existence of teamworking in a company is in itself fairly meaningless.

The perception by designers that they are more creative may in fact be a consequence of the increasingly stunted cultural ambitions and aspirations of the profession. The term design management contains a fundamental contradiction.

Design is about exploration and risk taking - as such it runs the risk of failure. Management, on the other hand is about control and predictability. Put the two together and design begins to lose out.

Design management literature has begun to seriously address how CAD can be implemented in such a way to curb creativity. Culverhouse (1995) proposes new practices to reduce design time and thus time-to-market, arguing that "it seems sensible to provide a knowledge constrained framework within which design may progress, given the need for management control of the design process".

One of the fundamental contradictions of this system is that increasing management control and predictability is at the cost of decreasing design's potential for innovation - the very reason for investing in it in the first place. Armstrong and Tomes (1996) have discussed this in terms of the increasing specification of briefs in design management. They conclude that in the process of making design a predictable tool of corporate planning "design will converge on a bland mediocrity and thus, incidentally, fail in the task which it was to accomplish for the corporation - the creation and location of new product markets".

Bill Hollins (1996) provides another view, consistent with this, that suggests that the very logic of British management based on short-term financial return combined with practices such as business process re-engineering, downsizing and benchmarking lead to product strategies based on small profitable niches rather than large market share and concludes that "many of the management actions and policies are making the organisations inflexible and unable to manage the design of their products and services into the future" (Hollins, 1996).

So, design management contains within it some inherent contradictions arising from the very logic of late capitalism which need to be considered more fully and empirically. As part of this we need to explore and define the diverse forms of design management. During a study to investigate the match between design management practice and design management education a series of interviews were undertaken in the summer of 1997 with individuals whose role included design management. The aim was to establish what their roles were as Design Managers, how they operated in their organisations and what knowledge and skills they required to operate effectively as Design Managers. These individuals worked either in Design Consultancies or as in-house Design Managers in major UK companies. The following case studies represents the findings for four of the interviewees.

Case study No. 1 Alison Fitch, Reebok U.K.

Background

Alison trained as a Fashion and Textile designer with a BA (Hons.) degree from Manchester Metropolitan University, between the years of 1987-1990. As a graduate her first job was with a fashion and textile importer. Although she initially began work here purely as a designer over the four years she was with the company she was promoted to a position where she took on many more responsibilities including design, sourcing and procurement, and manufacturing management roles. Before leaving to join the team at Reebok Alison found herself in sole charge of certain accounts and managing the entire design process from initial concept stages through to manufacture and production. It was during these years that she first learnt about the Fashion Design Industry as a business, she learnt to cope with people - both customers and suppliers, and she learnt about project management.

Alison first began work for Reebok in 1994 where she was appointed as Design Manager. She was in fact the first design manager there, and the role occurred as a result of restructuring. Previously such concerns had been the domain of the Senior Designer and the Product Development Manager, now this work has been divided between the Design Manager and the Sourcing Manager. It is the task of the Sourcing Manager to over see the Production and Manufacture side of the design development; visiting factories currently in operation; establishing possible new sites for production; dealing with the Human Resource problems related to the factories and their staff etc. Whereas the Design Managers concerns are entirely concerned with the design and production of the garments themselves.

Although Alison is constantly liaising with the Senior Brand manager, the decisions remain very much led by the design department, and future directions and briefs are devised by Alison and her team of designers. This is a key role for Alison as Design Manager.

The role and responsibility of the design manager

Alison's full job description and title is of Apparel, she is in charge of all design work that is carried out in-house in the U.K. Alison defined her role and responsibility as follows:

Design	*"My core role as Design Manager is being responsible for the ranges themselves and making design decisions"* *"Understanding the structure of the range and co-ordinating it."* • Developing a vision for new ranges • Working with the designers on trends and predictions
Management	• Managing the design and development of the ranges
Planning	• Production schedules - managing the design process to production hand over • Working with the sourcing manager to procure the samples and apparel production
Communication	• Presenting to the sales force • Working with marketing to develop the product support in the form of promotion and fashion graphics • Working as "an ambassador for design throughout the company"
Personnel	• Managing a team of eight designers • Dealing with Human Resource issues which occur such as Performance appraisal, job specification, and designer training and development.

Human resource management has become a major element of her work - "Since I started it has changed and a great deal of my job was involved in designing my own ranges whereas as the teams have been built up this has shifted and now I am basically in charge of keeping them motivated and ensuring they are doing what is required of them".

Skills required

Alison believes that a main reason why she has taken so well to this particular role has been her own family background and instinctive love of being with and working with people. "I was brought up in a pub and learnt to mix with an awful lot of different people from a very early age. This has helped a great deal as you have to deal with so many different types of people in this job". "Instinctively I love to be with people and as a result I don't find it a trauma to manage a design team and this is essential. You cannot ever feel insecure. You are handing design responsibility to a team of people and you need to have the confidence in them in order to make it work".

Design Skills: Within this particular role as Design Manager a background of practical design has proved to be essential - enabling Alison to fully appreciate the task in hand, and to plan accordingly, and also to enable her to design and to step in whenever necessary.

Design Awareness: Ability to visualise how a design on paper will look three dimensionally.

Interest in Design: A keen interest and aim to keep the ranges forever new and different is essential - especially as the very nature and structure of the ranges would make it very easy to lapse into a situation where very little is ever altered.

Design Vision: It is essential to be both creative and practical in the sense that the Design Manager must fit the design ideas into a commercial and realistic context.

Commercial Awareness: It is imperative it is to understand design work within its business and market context.

Personal skills and attributes: Equally as important as academic and practical design skills, initiative and instinctive personal skills also seem to be essential.

Patience: Alison rated this as one of the most essential skills for a Design Manager, as often finds herself incredibly frustrated when trying to get her own ideas and design suggestions recognised by others in the firm.

Persuasion: Ultimately it is Alison's job to present the design team's ranges of work to the sales force and this involves constant persuasion both visually and verbally. "The Design Manager needs both vision and persuasion". in the past the sales division have had a great deal of sway over the designs chosen for production. "Constantly battling to gain respect from a design aspect"...Alison has spent her years at Reebok trying to convince others that design is far more than arts and crafts and very highly directional and persuasive as a "selling tool".

Confidence: in both yourself and the design team: It is essential to personally believe that the work you are presenting is a good proposition: "There is nothing worse than presenting work that you yourself do not believe in".

Adaptability: The nature of this job entails a character that "can wear many different 'hats'". In addition to this Alison has had to adapt her own skills in line with the company; Reebok have a very strong culture, and Alison has been on a number of courses to develop a Reebok approach to such activities as interviews and presentations.

Polyphasic: An essential skill of design management for Alison is to keep a number of balls in the air at any given time - being able to juggle ideas and worries simultaneously and with relative ease. "You have got to be able to have 20 million things going on in your head at once - you have to be able to juggle".

Imagination: "Not only in terms of design but also in terms of how to use people to their best advantage".

What is design management?

For Alison design management involves..."Range building in a commercial market place; building into ranges; the correct and most appropriate use of colour and interesting fabric; creating the right silhouettes; devising price pointers; creating a range that is commercial enough to appeal to a wide range of customers. The Design Manager is employed in order to ensure that each range incorporates each of these elements regardless of who actually designed it."

Case study No. 2 Karin Ward, Prudential

Background

Having studied both 'O' Levels and 'A' Levels Karin received direct entry into a degree in Graphic Design. As a graduate Karin's first job was working for Waltham Forest as their in-house designer; this role taught her to be very independent as she was very much working alone and with all design elements. Having decided she needed commercial experience of her own Karin set up in her own business with a college friend. On leaving this business Karin went to work for MGM Cinemas as a Graphic Designer for their video literature and again was very much in sole charge - "There were ten people in the states doing what I alone was doing". From this job she then joined 'Crown House plc.' where she implemented a new corporate identity. After one year working there she joined the in house design team at the Stock Exchange until finally joining Prudential in 1989. Completing her transition from designer to manager of design.

When Karin first joined Prudential she worked within a separate design division. She headed this team, turning them into a unit that could service the company professionally, mainly working towards producing the marketing literature in response to the various divisions requirements... "Competing against all other design agencies and fighting against the prejudice that out-house was better than in house".

After a period of intensive restructuring in the company as a whole, the decision was made to out-source all design work, the design team therefore disbanded and Karin took on the role as Head of Design - a post she has now held for over a year.

Since January of 1996 the company has centralised into what is known as Prudential U.K. Since this time Prudential have created a bank and for various legal reasons this needs to be a separate operating arm. It is the Strategic Team which oversees the work of each of these 'arms', and this is where Karin's role as Design Manager sits. Much of her responsibility lies in the look of Prudential Assurance and the Bank.

The role of Design Manager

Karin's role entails a number of diverse, yet closely linked activities:

Design Policy Definition	• Considering the corporate identity of the company... "Clearly we have to use the identity correctly but what we are trying to do is to put the Prudential brand into a tangible form, a brand is often not a tangible thing, it is more of an emotive feeling about what you think of a company. What we are trying to do is put this down on paper and into our literature". • Co-ordinating the company sales literature... • Producing a Report and Communications Strategy... Part of Karin's role has been to work with the results of commissioned market research to produce a Communications Strategy for Prudential; within which she has clearly defined what the brand values are, what it is they are trying to achieve and a plan.
Managing the design process	• Selecting a preferred list of design consultancies Selecting those design consultancies appropriate for Prudential. "Ideally we will get agencies together like a family where they know each other and work together upon projects. Creating a feeling that they themselves are an extension of Prudential staff." • Devising a design brief and implementing a design process... "In effect what I have done is create a design management process which is very much tailored to this company specifically and operates in a company which is to some degree resistant to using a central resource - they much prefer to do it themselves and it is taking that into consideration". • Selecting an appropriate design solution... "If we were a company only selling a new product it is a much easier proposition. People often hold up BMW as an excellent example of Brand Management - though, with respect to BMW, they are selling cars which are a luxury purchase and a desirable purchase and which operate in one or two markets - they can sell the smallest to the largest BMW in the same way". • Over seeing the rolling-out of the new design work... Karin's role ends in the actual production of the design work.

| Communication- Selling Design | • Establishing the role of design in the company's success... Karin believes part of her role lies in convincing others within the company of the importance of design and the 'Added values' that it can offer... "The drive for design comes from the need to service your customer better and add value and to differentiate. Because there is nothing else in financial services to differentiate as they are all very much of a muchness". |
| | Karin believes that her main battle is getting her role in design management seen as necessary by the other members of staff at Prudential... "It is because it is design management and the company has never had design management and to them anything that comes out of this area they find hard to understand because they have never had to go through this process". |

Skills required

Design Awareness: "I do not necessarily think that you need to be a designer but you do need to have an appreciation. I think more than anything else it is communication and understanding what the company is trying to do with its literature and to ensure that it does that to the best of its ability".

Management skills: "An ability to instigate and manage process in a sympathetic way".

Persuasion: "Working within large organisations and working with people generally who do not have a real understanding of design and who actually see it as that unnecessary bit of frippery - you need to be able to help them to understand the value of it."

People management: "It is about handling people and delivering the message in a way that people do not feel threatened or patronised"... "A relationship with people is very important and an appreciation of where they are coming from; the skills are working with the major stake holders and the ability to bring them all together".

Interpersonal skills: "Possibly one of the most necessary elements of the job and this is where I feel designers often fall down. Some good designers find it hard to talk to those who do not understand what it is that they are trying to say - they get both frustrated and irritated and this comes across".

Patience: "Because it is a very frustrating job...certainly in my experience. Mainly because my role has not been in place before and there is an awful lot of resistance to 'another process'. The only way you can prove to them is to do it and this takes patience as the results are not immediate".

Flexible yet prepared to be firm: "Must be able to give when it is appropriate to give and be hard line when you need to be hard line, it is being able to identify where you need to put your foot down and where you need to stand back".

An eye for design: "I have an assistant working for me who is not a designer but she has a very good eye for design, so I can see how an individual can do the job without being a designer... although she does not possess practical design skills she has a very good eye for design and an appreciation of the process and an understanding of where creativity starts and stops and where pragmatism and logic comes in".

Natural Ability: "There are elements of natural ability which are essential and you cannot create - and these would be an eye for creativity and an eye for design; if you haven't got that, any amount of understanding processes and all that sort of thing is just not going to get anywhere"..."Clearly, when interviewing - you need to see something that gives you that comfort factor that they have an eye for design".

Definition of Design Management: "For me design is communication and as Design Manager you are the custodian of the visual communication to the customer".."Basically you are responsible for maintaining the consistency of the communication output to the customer"..."It is not just paper based - it is the entire question of how we communicate to our customer".

Case study No 3 Clare Newton, The Chase

Background

Clare began work as a Design Manager for 'The Chase' in 1996 and she is one of four Design Managers within the Consultancy, having originally studied for a degree in Design and Technology which involved a quite practical approach to design - focusing upon graphic and furniture design; with additional elements such as Silversmithing.

As a graduate she began work in a very small publishing company as a Mac operator and typesetter. Although she was involved in page layout it was "very crafted typesetting, rather than graphic design". Within this same company she was promoted to the role of Studio Manager where her role lay in over seeing the work on the printing press, along with the activities of the designers. She was responsible for the work of four designers, two of which were juniors, and a copywriter - making sure schedules were devised, work was completed on time and to a high standard.

From this role she then joined a company called 'ADS' in Piccadilly which specialised in Telemarketing but which had an art work division. Clare worked as 'Account Handler' although she relied on her own skills as this post was the first of its kind within the company. A far more sound grounding in this role was gained when she began work at U.A. Simmons - an Advertising agency, also in Manchester. Working within a far more structured design studio she was taught a number of skills from handling clients effectively to general administration and paperwork skills. Clare then moved to 'The Chase' as Design Manager.

A total of twenty seven people work at 'The Chase' and these people are divided into three teams each comprised of one or two Senior Designers, a Creative Director, Junior Designers, a Team Leader and a Design Manager (of which there are four in total).

The teams themselves are client based and work with specific clients; Alison's team being primarily responsible for the Co-op bank, along with a few very much smaller clients.

The role and responsibility of the Design Manager

Clare describes her role as "Everything but Design…" "The friendly role as well as the financial role"… "the management side of the Designer / client relationship".

As a Design Manager Clare's main responsibilities includes:

Planning	Devising Production schedules and also weekly Production Management structure.
Finance	"A main responsibility lies in the invoicing and the chasing, particularly if an account is in trouble in any way".
Communication	With the client, the designers, and also suppliers. A large amount of time as Design Manager is spent liaising and communicating with the client - "Taking them out to lunch and ensuring everyone is happy with how the design work is progressing and dealing with any problems".

Skills required.

Communication: "Definitely communication, definitely getting on with people".

Persuasion and sales ability: To be a successful Design Manager Clare believes it is essential that you have such abilities as…"You are, in effect, selling the company".

Interpersonal Skills: "Need someone who can talk on many different levels. Build up relationships - clients need to feel they can approach you, rather than the Director when they are not happy". Intuitive skills are also essential which Clare believes - "You either have these skills or you haven't" . She also stated it to be essential to be a people's person and to fit into the team. "If the design team disliked you it would prove to be incredibly hard".

Articulate: "The Strong Arm of the Designer". In many ways Clare believes that the Design Manager's role is to represent the strong side of the designer who is not afraid to talk about schedules and costings and "all of the practical and realistic elements that designers are often loathe to talk about".

Organisation: Clare believes this to be imperative in her role as design manager..."I tend to have a to do list two pages long which I draw up every night to ensure every thing that should be done is." ... "because when you are working with creatives they have little desire to be organised".

Visual Awareness: Clare believes that more important than practical design skills is visual awareness... "They need to be aware. It is a real asset if the Design Manager is very aware of design and actually has done it in the past". However, she also stressed that there is a danger in being too confident in your own design skills... "It can be difficult if you are actually a designer going into design management because you cannot help but get quite involved and have your own personnel opinion on something - and this can cause some anxiety".

Business Negotiation Skills / Commercial Awareness: "Business ideas are important as it gives you credence and makes people listen to you".

Financial Awareness: Clare stressed the importance that the Design Manager must be financially aware and willing to work with figures.

What is design management?

Clare's own definition of design management was short and simple - "Everything but Design". Within 'The Chase' design management is very much concerned with issues of PR and personnel, constant liaising and communication with the client is a core part of Clare's role as Design Manager. Whereas the design itself is the concern of the designers, the Design Manager only tends to intervene to ensure that the designs are realistic in terms of cost and appropriate in terms of the client's aims for the design.

Case study No. 4 Peter Hampell, Imagination

Background

Peter holds a degree in modern languages, he spent the first twelve years of his career in advertising. Working first on the account handling side in the London office of an American advertising agency, then joining Saatchi and Saatchi as an Account Director working for clients such as Kellogg's, The Sunday Express, The Sunday Telegraph, Castlemain XXXX, Holsten Pills and Cadbury's. From this Peter then joined Imagination as Client Services Director.

Over two years, Imagination was repositioned under the broad concept and heading of the "brand experience". This was partly because senior management believed that they could no longer operate profitably in the long term within their specialist marketplace and needed to devise a form of positioning which would allow them to compete in a marketplace beyond their core businesses of exhibitions, product launches and conferences.

The main aim of Imagination is to persuade others that they need far more than just advertising strategy and that it must form a far wider picture which encompasses a complementary brand experience strategy with a brand strategy... "..encouraging a consistency in corporate voice and corporate communication across all that they do". Whereas most view design as a very linear and narrow sphere…. "It is all about the entirety of the designed elements; what the salesman says, what colour the van is, what type the van is, the receptionist, what things look like etc. All these elements are essential and this Holistic approach is our overriding aim". "Moving into the strategic formulation rather than staying at the product end and what that meant was strategic relationships and not project based execution - devising long term partnerships" ...With this shift in emphasis came a realisation that a different type of Project Manager was needed; "People who could not only manage and co-ordinate a process, handle a client, understand a brief and translate it into something the creative could get excited about and work with and then co-ordinate a production process - on time and on budget. We needed people who could do all of this plus understand a brand, understand a brand strategy, and understand how to develop a partnership and long term relationship".

To facilitate this approach they developed three major prongs to the client-facing unit:

- *The Design Director* responsible for the artwork and design work itself - ensuring that the work produced is in line with the client's initial aims and specification:
- *The Account Director* with overall responsibility for the Design process and relationship with the client.
- *The Planning Director* responsible for strategic input, marketing context and consumer context, along with information about the brand: Basically all marketing and strategic issues... "Such a role is in fact unique in design work. The aim is to establish the client's whole operating strategy and operating environment and learn from the client where they should position themselves etc..." It is quite a creative role in a sense and one which hopefully provides some insight, and helps in defining the crucial elements such as target audience, competitive context, what the brand itself is about and what the product can achieve". Because they provide direct planning input they are fairly fundamental to the creative product.

Peter's roles and responsibility

Peter is the Client Services Director and as such sees this as design management, his roles and responsibilities were described as follows:

Client interface	"I am responsible for the client relations, co-ordinating the Imagination resource to deliver the design work against the set objectives". It is Peter and the Marketing Director who initially meet with the Client… "The marketing director sells the company and I explain how the company can deliver such promises in terms of design; describing the processes involved, how they will be handled, who will be responsible for them and their relationship with the Consultancy, along with discussing the more practical issues of contracts, numeration deals etc. but most importantly determining with the client their actual design aims and current problem" "Interfacing with the client at a Senior level". As Client Services Director Peter is responsible for building the client - Consultancy relationship and identifying the projects - depending upon what the clients communication requirements are.
Client marketing strategy formulation	"With the clients marketing strategy in mind I then advise them on how this 'design tool box' can help to deliver and meet those marketing objectives". When dealing with clients it is Peter's role to assess their situation and to determine, what issues are facing them, and how they interact with other agencies. However, although this role is carried out externally in terms contracting clients and making them "Understand what is Brand and why is brand experience important ". This also has to be done within the company to all staff.. "my job is internal monitoring and external monitoring and both are equally important."
Management and communication	With the changes in company structure Peter's current role has become…"Clarifying these new roles both internally and externally, clarifying the process that are involved and making sure they are all doing it properly - implementing processes and skills

Skills required

Design Understanding: "Although someone need not be a designer they would still require an intuitive understanding of design and an excitement by creativity". An interest in creativity and its effect on other people - "This is what should excite anybody who works in the creative business." …. "The skill is in accessing the specialist resources and not necessarily having specialist knowledge. Knowing when to draw people in and how to use people is essential".

Commercial Awareness: "An interest and understanding of the ways in which business works, marketing strategies are put together, and how brands operate." An understanding of the general issues and challenges that confronts marketers in whatever industry they are working in - on a general but also a specific level... "You have got to understand the marketing business context within which you are working".

People Management: "It is a general interest in the psychology aspect: you have got to be interested in people, and not just from a consumer perspective. Your job needs you to be excited by as well as excite a whole lot of different people within the whole process. If this comes across as anything but genuine you are rumbled fairly quickly. You have got to be skilled in motivating people".

Client Facing Skills: "The client needs to feel absolutely confident that you are in control of a process".

Intuition: "You need to be naturally able to spot and draw out the issues when meeting the client; you also must know which questions to ask and when to challenge a client and how to rationalise your ideas and arguments".

Interpersonal skills: "You have got to be able to listen, and have a good antenna to recognise sensitivities and politics etc. You cannot bulldoze your way through and need to be very sensitive to certain issues. A great deal of marketing is common sense - the intuitive and clever bits are knowing how to manage people because this is where the satisfaction comes from".

Adaptable and Flexible: "There is a need to be infinitely flexible in the creative media they work within because they should be interested in it - a jack of all trades ... You need enough understanding of all of the creative disciplines to be able to talk sensibly and with authority to a client who knows less than you do".

Peter's own belief: is that ideally Design Managers should be recruited and their abilities built up gradually. He maintained very strongly that a great deal of the skills needed are about personality, personal skills and intuition.

Definition of Design Management: "I see design management as more of a client role. I see it more as someone who is trying to co-ordinate what traditionally falls into design areas and different design disciplines from the clients perspective and ideally working very closely with the marketing director as design should be as much about marketing and communication as it is about physical structure and two dimensional graphics".

Towards a typology

While only based on four case studies, we can propose a typology of Design Managers that defines very different and distinctive activities:

Creative team manager	Based on the Reebok case, the emphasis is on providing design direction and focus, managing a team of designers, and providing an interface with other corporate activities such as marketing and sales. Key skills are design, design leadership and personnel management.
Design procurement manager	Based on the Prudential case, the emphasis in on managing the relationship with external design consultants, defining corporate design policy and negotiating solutions with consultants. Key skills are design awareness, process management and general management.
Account manager	Based on The Chase case, this is characterised as "everything but design", with a focus on financial control and, appropriately, chasing. Key skills are financial management, negotiation skills.
Marketing manager	Based on the Imagination case, design is seen as a sub-set of marketing with the emphasis on marketing strategy and project management. Key skills are marketing, strategic management and communications.

To these four categories, others could be added. Recent research by Chamberlain, Roddis and Press (1998) views design in terms of 'entrepreneurial mobilisation' which we could characterise as 'process manager'. In this case the designer manages and pieces together all parts of the NPD system.

These four case studies illustrate that the role of Design Manager is multi-faceted, and currently dependent on the individuals' own interpretation of the their job. It is perhaps a term applied to them rather than a profession. These individuals are responding to the needs of their business and the contribution they can make to enabling design to be used effectively. The question is: will the next generation of design and design management graduates be able to do the same? Before the curriculum of courses such as BA in Design Management become too embedded in the system, it is time we question the content, the orientation and the philosophy which arises from them.

This raises the issue of the vision that is propelling design education into the next century. While the 'romantic hero' narrative that justified design education for much of this century was in need of reform, is the 'manager' model the only alternative? Is 'design thinking' just a form of 'management thinking'? Self-identity has perhaps never been a strong point of designers and design educators. Design management offered an identity appropriate to the culture of the 1980s, but today's culture, and the problems facing it, is different. The need for sustainable industry, the challenges of new technologies, issues of usability, the desperate need to rebuild our impoverished and ecologically scarred inner cities, to draw understandable maps of the crowded landscape of information - all these things demand design thinking, design imagination and design in the service of citizens.

The contradictions identified earlier can be tackled if we explore the notion of design citizenship, and embed that within design courses. Existing design manage-ment courses, especially at postgraduate level, should then define themselves more in terms of the specialist fields of management that determine design processes - creative team management, design procurement management, etc.

Has the term 'design management' outlived its usefulness? Is it time to bury it?

REFERENCES

Armstrong, P. and Tomes, A., 1996, Art and Accountability: the Languages of Design and Managerial Control. In *Accounting Auditing and Accountability Journal,* (MCB University Press Ltd) Vol. 9, No 5., pp. 114-125.

Chamberlain, P., Roddis, J. and Press, M., 1998, *Good Vibrations*, Submitted paper, DRS Conference, Quantum Leap: Managing New Product Innovation, UCE, Birmingham, September 1998.

Cooper, R. and Press, M., 1995, *The Design Agenda: a guide to successful design management*, (Chichester: John Wiley & Sons).

Culverhouse, P.F., 1995, Constraining Designers and Their CAD Tools. In *Design Studies,* Vol. 16, No. 1, Jan., pp. 81-101.

Fisher, T., 1996, *Creativity and Teamworking*, conference paper, Design Management Institute Conference, Barcelona, November 1996.

Fitch, R., 1988, Impact of Design on the Retail Landscape. In Gorb P., (ed.), 1988, *Design Talks,* (London: Design Council), p. 154.

Gorb, P., 1988, *Design Talks*, (London: Design Council).

Gorb, P., 1990, *Design management: papers from the London Business School*, (London: Architecture, Design & Technology Press).

Hollins, B., 1996, *Are Current Management.Practices Damaging Long-term Design Management Effectiveness?* In proceedings of 8th the International Forum on Design Management Research and Education, Barcelona, November 1996.

Oakley, M., (ed.), 1990, *Design Management: a handbook of issues and methods*, (Oxford: Blackwell).

Powell, W., 1995, Design: no good sitting on the side lines. In *Co-Design Journal*, issue 3, June.

Roy, R. and Potter, S., 1993, The commercial impacts of investment in design. In *Design Studies,* Vol. 14, No 2, (April): pp.171-95.

Service, L.M., Hart, S.J. and Baker, M.J., 1989, *Profit by Design,* (Scotland: The Design Council).

Wickens, P.D., 1995, *The Ascendant Organisation: Combining Commitment and Control for Long-term,* (Basingstoke: Macmillan).

ACKNOWLEDGMENTS

The authors wish to thank all those Design Managers who gave their insights and time to the researchers. Also Kate Shepherd final year undergraduate in Design Studies (1997) at Salford University, for her research which contributed to this paper and Chris Rust of Sheffield Hallam University for his critical comments on a draft of this paper.

Building Electronic Totems to Manage Automotive Concept Development

Andrew Fentem and Angela Dumas

In collaboration with large manufacturing organisations we have developed tools and techniques to improve the communication and management of the strategic knowledge essential to product development. This set of techniques, known collectively as Totemics, can be used to explore product strategies, potential new product concepts, brand identities, and market positions. The process is initiated by using visual metaphors to facilitate cross-functional conversations around the values to be associated with new product concepts. During this exploration images, objects and words are arranged to create a map describing potential product strategies. This 'totem map' can be used to transfer values and tacit knowledge between different domains in the organisation, such as engineering and marketing.

Support for this process is provided by a combination of facilitation and a software system that supports linkages between the totem map and the artefacts of more detailed product development work that takes place across the organisation, in any medium. This paper provides the background to, and description of, a case study of a collaboration with the technical strategy group of a large European automotive product manufacturer where we have deployed these techniques. The paper describes the system implementation strategy and how multimedia tools are being used to integrate the map with extant styling, engineering, and market knowledge. The trials of this process are proving to be effective, and this technique is being implemented in other new concept development teams and other organisations, to facilitate the design of services as well as physical products.

INTRODUCTION

A major problem for knowledge intensive organisations, particularly those involved in design and manufacturing, is the increasing specialisation of knowledge communities or functions. This specialisation causes cross-functional communication problems, because each function has its own language and perspective. Increasing isolation results in cross-functional conversations becoming more impoverished as employees identify less with an organisation's culture, brand identity and values. One of the effects of this is that product coherence will become even harder to maintain. The use of electronic communication systems will exacerbate this problem through increasing the isolation of these teams. At present these systems do not facilitate the transfer of the tacit knowledge that is crucial to any organisational learning process (Nonaka, 1994).

It is thought that these early problems could be overcome if a shared vision (Senge, 1990) of the product identity is communicated at the beginning of the design process. In order to communicate this vision, non-codified or tacit knowledge (Polanyi, 1966; Nonaka and Takeuchi, 1995) must be converted into explicit knowledge, before it is transmitted, through the use of metaphor say, or perhaps communicated through richer, more 'experiential' multimedia environments.

Research into the role of visual, object and verbal metaphors in communication and thought (Ortony, 1979; Lakoff and Johnson, 1980), and the ways in which cognitive mapping techniques can assist strategic management (Huff, 1990), has informed the development of a technique known as Totemics. This paper describes how this technique assists groups and management teams in creating dialogue and exploratory conversations addressing a wide range of product strategy issues and facilitates the transfer of crucial tacit knowledge across domains and functional barriers. The participants experience how metaphoric explorations can work to elicit the tacit knowledge held within and between a group, and they learn how to create an interface, or a bridge, between the tacit and the explicit knowledge of the group. The paper describes how this interface is created through the plotting of visual and verbal data in a map that describes the values relevant to the product issues under consideration.

Multimedia tools are described that have been developed with the industrial collaborators involved in this project. Development of these tools was initiated in order to sustain the conversations across the participating organisations. The tools present the maps in a user-friendly interactive fashion and provide support to a team in their on-going explorations of this strategic knowledge and its association to more detailed descriptions of the new product designs.

THEORETICAL BACKGROUND

Global competition is such that product and service development is now of much more central strategic concern. Strategic benefit can therefore be gained through the integration and coherence of the organisational, market and new product strategies. This integration process can be achieved by the generation and sharing of a common 'vision' (Senge, 1990). This vision, a form of strategic knowledge positions and guides future work, enabling all functions to pull in the same direction, while operating reflectively and creatively. This shared vision can be reinforced through familiarity with a shared set of values that underpin all of the activities in the organisation. These values are usually tacit. Making them more explicit can encourage individuals and teams to be confident that any innovations that they generate will be broadly in-line with the activities of other teams and individuals, and the future direction of the entire organisation.

Metaphors are operating most of the time in language (Lakoff and Johnson, 1980), being used to transport and translate concepts across different situations and contexts. This function can reinforced by adding images or objects that visually support the point being made. In whatever form, metaphor can prove useful in stimulating creative thinking and in the transfer of knowledge across domains in many different contexts. For example, metaphor can be used to provide vision in the management of new product development processes (Clark and Fujimoto, 1990; Nonaka, 1991; Dumas, 1994). Similarly, the totemics process uses visual metaphors to create a 'totem' that provides access to the multidisciplinary tacit strategic knowledge involved in new product creation, and the transfer and integration of the knowledge between domains.

Another familiar method that can be used to integrate strategic knowledge is mapping. Maps of all sorts have been shown to be particularly useful in strategic management thinking in organisations (Huff, 1990; Weick, 1990) because they act as a representation of a perspective which, through being challenged, can enable learning. Maps help to organise discontinuities and contradictions and gradually bring pattern, order and sense to confusing situations. They may also be appreciated in different ways by different groups, while at the same time possessing coincident boundaries that are understood by all groups. In this way maps act as a type of 'boundary object' (Star, 1993).

The totemics technique takes advantage of the properties of these two familiar mechanisms in the generation of semi-structured maps of words and images to form rich representations of product strategy knowledge. These maps that act rather like totems (Lévi-Strauss, 1966), implicitly referring to the participating group's shared values, tacit knowledge and language games. They are not intended to be completely deterministic and unambiguous, but are incomplete and partially indeterminate, indicating the terrain of what is known as well as that which is not. They are not necessarily supposed to be accessible to someone who is not part of the participating group who constructed the interpretive structure, but exist to assist in the management of the group's strategic vision, enabling it to maintain coherence and innovation across all aspects of its operations.

TOTEMICS CASE STUDY

Throughout this research project, a form of action research (Clark, 1972), a great deal of emphasis has been placed upon the co-operative relationship between the researchers and the collaborating industrial organisations. The refinement of the knowledge elicitation techniques, and the development of computer-based tools to support the operationalisation of this knowledge, is a highly participatory project. This co-operation has enabled the introduction of new tools and techniques into the organisation in such a way that both groups can explore and learn how they can be used. This ultimately leads the collaborating organisation to develop a good understanding and sense of ownership of the processes, tools and representations, which is a particularly important consideration if designers are among the target users (Mitchell, 1993).

One of our main industrial collaborations is a long-term project taking place within a large European car manufacturing organisation. The totemics technique is being deployed across the organisation to enhance the exploration and communication of product strategies. Its implementation is assisting cross-functional development teams to develop visual and verbal language that will enable them to develop their ideas, create a clear product strategy, sustain this dialogue within the organisation over the long-term, and provide linkages to their more formal and established product management processes.

The initial aim of this intervention was to aid the technical strategists to create a vision against which they could check when taking critical design decisions before vehicles are built, and guide the formation of the more qualitative attributes of the product. This vision would be in addition to the quantitative analytical representations of product plans and specifications already used to gauge whether a specific design will meet certain general financial or performance criteria. It was hoped that this vision would act as a bridge between the quantitative information used in some departments and the representations of qualitative information used in the marketing and design departments. These representations, often called 'mood boards', are stimulating but often unstructured collections of images that attempt to metaphorically capture the qualitative aspects of the current product strategy.

During the facilitated stage of the process, the participants discussed the values that they associated with a very large range of objects and images, and those considered to metaphorically represent potential aspects of product strategy were selected. For example, a picture of a high-tech modern steel building such as the Lloyd's Building in London, might be chosen to represent 'clever engineering' as a desirable aspect of the new product development strategy. This process exposed the individual participants' perspectives both to themselves and to the other members of the group. These images and appropriate annotations were then positioned in a two-dimensional space, depending upon how the salient values of these images related to one another. In this way the group mapped out a terrain constituted by the differences between the images, expressing the range of different product strategies open to them. A typical such a map is shown in Figure 1, where for instance the top right hand corner could be understood to signify future product strategy.

The use of this map is being promoted across the organisation. The techniques are propagating slowly across the different departments in the company, with no intervention from ourselves, and are being used to form the basis of focus groups with consumers, and communicate with external groups such as advertising agencies.

MULTIMEDIA TOTEMS FOR COMMUNICATION OF PRODUCT STRATEGY ACROSS DISTRIBUTED TEAMS

The new product development projects in an automotive manufacturing organisation are long and complex, and the development teams are large and often geographically distributed. These factors can make it harder to sustain the strategic conversation when the more intensely facilitated stage of totemics is over. It is important that the participants communicate the product strategy map outside of the immediate participating group, refine it, and include product development detail into the product visions over time.

Figure 1 An example of a typical totem map

These requirements, together with the existence of very large quantities of information in many different media that need to be managed and integrated in this type of process, suggest that computer-based tools could help to sustain and stimulate the conversations.

Such tools do, however, need to be of a quite different nature to the type of information systems normally found in large bureaucratic organisations. During a previous collaboration with a large manufacturing company it was recognised that conventional Product Data Management systems, that are often used to manage engineering and design data, are unsuitable for propagating this type of information. This is because the structure of such information systems is too rigid. One of the main advantages of employing maps and metaphors to communicate qualitative design-oriented information is that they are tolerant of ambiguity and vagueness or indeterminacy, whereas conventional database systems are not; by their very nature they demand that the artefacts stored in them are explicitly categorised. In contrast, characteristics such as exploration, stimulation, and ambiguity are crucial to the success of a dialogue such as totemics in facilitating a design process.

In terms of actual requirements, the support system needs to enable the potential users (from design, engineering, technical strategy, and marketing) to manage their activities with some reference to the totem map. It needs to stimulate them to enhance, communicate and reflect upon the evolving spatial representation(s) of shared strategic understanding, and integrate this strategic knowledge with existing artefacts, processes, and domain knowledge. Through use in this way the maps they will gradually acquire more value.

As already stated, the system is designed to *enable* the new process within the organisation. An implication of this novelty is that normal requirements capture techniques for development of the system can not be used because there are no extant processes to analyse and model. Development of the system is therefore taking place according to an evolutionary growth model which has been used for the development of systems to support collective creativity (Fischer and Nakakoji, 1997). This process is iterative, consisting of three stages; the first stage is 'seeding' in which the system is sown with an initial knowledge structure; the second stage is 'evolutionary growth' in which the seed grows through interaction with the users and is appended with new information; the third stage is 're-seeding' in which the knowledge structure is reorganised to make it more manageable and improve the computational support. This development process is suitable because it recognises the dynamic between the evolution of the knowledge representations within the system, the evolution of the participants, and the evolution of the process. There are two important prerequisites for this evolution-through-use to take place; the system needs to be engaging and stimulating enough to initiate early interactions; and the map and the system need attain a 'critical mass' (Saffo, 1996) of utility, usability, and availability as quickly as possible.

The initial multimedia seed system used was a 'visual spreadsheet' system simply implemented in *Microsoft Excel 97*. This application was chosen because the system of large, active, and flexible two-dimensional sheets met our initial requirements for the space in which to build maps. Also spreadsheets allow users to easily place all sorts of multimedia objects such as graphics and sounds on a sheet, and position or hyperlink them to indicate the relationships between them. The sheets tolerate ambiguous relationships between the objects that constitute the content, thereby facilitating the gradual emergence of order and categorisation of new knowledge. Also, because the spreadsheet application was already widely used within the organisation, the set-up costs were minimal, and it could potentially be integrated with existing data from other phases of the product development process.

The spreadsheet system was seeded with both the visual and textual information from the facilitated totemics group process, including the maps, together with examples of tactics for structuring the existing information. After seeding, the system was used and new content added to the system, the users being free to organise their contribution in ways that they felt would best communicate its relevance to the rest of the team. The participants appended the totem map with new objects such as images, using a mixture of spatial arrangement and hypertext linking to relate new objects to the rest of the content. The familiarity of the spreadsheet application meant that the participants could easily add new content and even add their own additional rudimentary functionality to the system using macros. In this way connections developed between the strategic knowledge and the low-level detailed decisions, hence documenting of some of the normally tacit and more qualitative aspects of the new product development decision making process.

The spreadsheet system quite quickly became computationally unwieldy due to the fact that these systems are not designed to cope with large quantities of multimedia information, so the system was re-seeded by converting the contents of the visual spreadsheet system into a web site which enabled the knowledge associated with the maps to be accessed and managed more easily.

As an information resource, this hypertext solution has proved successful. There are however some problems with it as regards the development of an evolving information structure; there is little potential for indeterminacy of relationships between objects as all new associations have to be in the form of static explicit links and it is not possible to move objects around in the space, so as to gradually arrange the content; it is also not easy enough for novices to add content. An additional problem is that as the quantity of content increases, it becomes difficult to gain an overview of the contents. This lack of overview makes it hard to see how the product strategy knowledge is related to more detailed design knowledge, impeding the exploration of detailed knowledge without losing the wider context.

In our case study, a hypertext hierarchy of flexible spaces was observed to evolve in the visual spreadsheet system. This structure is similar to that of spatial hypertext systems (Marshall and Shipman III, 1995) that have been designed to assist knowledge management across workgroups, intended for contexts such as ours 'in which the domain structure is not well understood at the outset, or changes during the course of a task'. The systems that will form the next seed structures are based on this idea, consisting of a totem map space in which objects (such as text and images), and links to objects (such as images, documents, World Wide Web pages, and drawings) can be added and arranged as is deemed appropriate by the user(s). Objects in each space can also be links to other similar spaces in the open hierarchy of spaces. Within the spaces it is possible to magnify local detail in selected areas of a map, and perform analysis of the attributes of those objects.

CONCLUSIONS

As organisations become more knowledge intensive, they will need more flexible, visual and intuitive means of interacting with business process information, and will demand systems capable of tolerating and supporting the ambiguities and contradictions within our conversations. In totemics there is potential for nurturing the evolution of new kinds of abstract information spaces, 'digital totems', that are designed to accommodate these new requirements. Like the totems of other cultures, they visually allude to the stories and values that constitute the identity of the group and help them to behave coherently during the varied activities of new product development processes.

REFERENCES

Clark, K. and Fujimoto, T., 1990, The Power of Product Integrity, In *Harvard Business Review,* **68**, pp. 107-118.

Dumas, A., 1994, Building Power of Product Integrity. In *Harvard Business Review.*

Clark, P. A., 1972, *Action Research and Organisational Change*, (London: Harper Row).

Dumas, A., 1994, Totems: Metaphor Making in Product Development. In *Design Management Journal,* **5**, pp.71-82.

Fischer, G. and Nakakoji, K., 1997, Computational Environments Supporting Creativity in the Context of Lifelong Learning and Design. In *Knowledge-Based Systems,* **10**, pp.21-28.

Huff, A. S., 1990, *Mapping Strategic Thought*, (Chichester: John Wiley and Sons,), pp. 11-49.

Lakoff, G. and Johnson, M., 1980, *Metaphors We Live By*, (Chicago: University of Chicago Press).
Lévi-Strauss, C., 1966, *The Savage Mind,* (London: Weidenfeld and Nicholson).
Marshall, C. C. and Shipman III, F. M., 1995, Spatial Hypertext: Designing for Change. In *Communications of the ACM,* **38**, pp. 88-97.
Mitchell, C. T., 1993, *Defining Design,* (New York: Van Nostrand Reinhold).
Nonaka, I., 1991, The Knowledge Creating Company. *Harvard Business Review,* Vol. 69, No. 6, pp. 96-104.
Nonaka, I., 1994, A Dynamic Theory of Organisational Knowledge Creation. *Organizational Science,* **5**, pp.14-37.
Nonaka, I. and Takeuchi, H., 1995, *The Knowledge Creating Company,* (Oxford: University Press, Oxford).
Ortony, A., (Ed.) 1979, *Metaphor and Thought,* (London: Cambridge University Press).
Polanyi, M., 1966, *The Tacit Dimension,* (Gloucester, Mass: Peter Smith).
Saffo, P., 1996, In *Bringing Design to Software,* Edited by Winograd, T. (New York: Addison-Wesley), pp. 87-104.
Senge, P. M., 1990, *The Fifth Discipline: The Art & Practice of The Learning Organization,* (London, England: Century Business).
Star, S. L., 1993, In *CSCW: Cooperation or Conflict?* edited by Easterbrook, S. (London UK: Springer-Verlag), pp. 93-106.
Weick, K. E., 1990, In *Mapping Strategic Thought,* edited by Huff, A. S. (Chichester: John Wiley and Sons), pp. 1-10.

ACKNOWLEDGEMENTS

This project is funded by the EPSRC in the UK, grant reference K/48228. We would like to thank Janet McDonnell in the Department of Computer Science at University College London, Eleanor Toye at the University of Cambridge, and all of the organisations that have collaborated with this work.

Design Orientation in New Product Development

J. Gotzsch,

This working paper reviews literature and research in industrial design. Industrial design is more user-focused than technical product development and is also influenced by aspects such as cultural preferences for aesthetics and the products' emotional value. These aspects are not enough developed in 'new product development' literature. At the end of this working paper further research in this field is proposed.

INTRODUCTION

Prior research shows that design is important to a company and to the success of its new product development. This paper argues that this importance means that the management of the design process, especially the management of the inclusion of the emotional factors into the design process, deserves more attention.

The first part of this paper describes the context. This section distinguishes product development, research and development (R&D), and art. The contributions of design in the new product development process are presented in a model.

The second part presents a model that describes the different factors and drivers for new product design. This model is used in the third part as a framework to summarise the literature on design.

The fourth part of the paper identifies some critical gaps in the knowledge about the management of design as part of the new product development process. One of these gaps is a lack of understanding on how to manage effectively the inclusion of the emotional and symbolic value of a product in the design process. Many firms have difficulty in managing this aspect of product development. To bridge this gap, the paper proposes research on the methods employed by successful design-oriented companies. An understanding of these methods would provide benefits both for other firms and for the development of effective management educational programs.

Research context

This working paper reviews and summarises design-oriented product development literature. Only design-oriented articles and books containing key words such as "design," "industrial design," and "product design" have been used and will be referred to as design-oriented literature. The terms industrial design, product design, and design, used in this paper, refer to exactly the same activity. Product design is only one aspect of design management. Others are packaging design, graphic design, multi-media design, retail design and architecture. In this text we focus on product design.

Well explored research close to the field of design-oriented literature is the new product development (NPD) literature. NPD studies investigate the best practices in innovation. However the extensive NDP literature has been scarcely used in this paper in order to keep a clear view on the design-oriented literature. While analysing a selection of design-oriented literature we find literature describing "Psychological responses to design" (Crozier, 1994), "Managerial attitudes towards design" (Hart and Service, 1988) and "Why design is difficult to manage?" (Dumas and Whitfield, 1989). These aspects and others, such as the use of product aesthetics, and the symbolic value of products or cultural preferences, are not developed in NPD literature.

Some institutes have focused on design-oriented research. Studies concerning design management and the efficiency of design have been made by the London Business School, the London Design Council and the Design Innovation Group (DIG). The Boston based Design Management Institute organises international design conferences and publishes the Design Management Journal. Research on industrial design is also undertaken at the Delft University of Technology, Faculty of Industrial Design.

However this paper shows that there still is a need for greater understanding on how to include design in the new product development process.

The design contribution model

The following overview of existing literature makes clear why design is important for a company. A visual model shows the relative contribution of design at the different stages in the NPD process.

Design has influenced a large variety of aspects in the company and for the consumer. An exact definition of design is not made here, but the most important elements of design are the following. Design is used for products of different price ranges. The seduction, pleasure and surprise elements are essential. Design gives personality to a product, and that is why we like the product. The designers' passion and care for detail are reflected in the product quality; and design is very much user-focused.

Peters (1994) illustrates this with a list of quotations on design, for example: "Design is ... total consistency, a design sense that pervades every single thing a corporation does, ... the product of an organisation that deeply honours playfulness and creativity, ... living (literally) with customers, ... a pleasant surprise ..."

Kotler and Rath (1984) define design's objective as "to create high satisfaction for target consumers and profits for the company." They propose the term "design mix," consisting of performance, quality, durability, appearance and cost. For them effective design uses these performance, quality, durability, appearance variables to create products at a cost affordable for the target market.

Walsh *et al.,* (1988, 1992) separate the influence of design in price and non-price factors. For example, design can have influence on the cost price by reducing the manufacturing costs and on life cycle cost by influencing the costs of use and maintenance. Product-related non-price factors concern product quality aspects, such as product performance, uniqueness, appearance, reliability, durability, safety and ease of use. Company related non-price factors concern company image, sales-promotion, delivery time (the product is designed for ease of development and to meet delivery schedules) and after-sales service (the product is designed for easy repair).

Roy, Walker and Cross (1987) mention (in Walsh, Roy and Bruce, 1988) that different design aspects are important at different moments of the product's life. Before and during purchase the perceived qualities, such as advertised performance and appearance, first impressions of performance, company image, price and test results of the product are important. These are called brochure and showroom characteristics. During initial use other aspects such as the actual product performance, ease of use and safety become important, the performance characteristics. And for the long-term, reliability, ease of maintenance, durability and running cost are important. These aspects are called the value characteristics.

Trueman (1998) proposes a four-level model to classify the attributes of design. She uses this model to audit companies in order to determine for which aspects a company primarily uses design. In this model design can have influence on a company at four different levels, namely value, image, process and production. At the value level, product styling, aesthetics, quality, standards and added value are important. At the image level, product differentiation, diversification, product identity, brand creation, corporate identity and corporate culture are mentioned. At the process level, the contribution of design is recognised in its capacity to update products, to generate, communicate, interpret and integrate new ideas, to function as interface (between managers, project team, production and customers) and to promote, advertise products. And at the production level, design can help to reduce complexity, production costs and production time, to use new technology and materials, to recycle products and materials.

Lorentz (1990) highlights the different process qualities of design for the creation of new product ideas. The designer puts his/her imagination in the service of marketing to win the consumer's favour and to find significant product differentiation. Moreover the designer facilitates the design process as he/she co-ordinates, evaluates and executes. In detail Lorentz mentions the designer's characteristics, such as drawing skills (that facilitate the visualising of ideas), communication skills by presenting drawings, creativity, designer's natural suspicion towards traditional solutions, and designer's capacities to synthesise very different external and internal aspects. He/she recognises that design is no longer an option or a strategic choice of marketing, but an essential function within the company. Marketing and design are indispensable to adapt the product to a target group that needs to be defined more and more precisely.

Different and independent studies have been made to measure the efficiency of design, as management is often uncertain about the efficiency of design. The main objective of these studies was to obtain clear figures of the economic value of design. These studies show a positive correlation between the use of design and good business performance. However, the pure effect of design is difficult to measure and isolate (Cote-Colisson *et al.,* 1995). Business success depends on many other factors, such as launch strategy and quality of production. In general good design helps, but design alone is not enough to make a company flourish. The French design performance study by Cote-Colisson *et al.,* (1995) found, apart from financial aspects, the following general aspects concerning the influence of design: design gives a competitive advantage as it provides product "differentiation" with a "rigorous product conception process," good design improves "technical performances and production costs". The value of innovation is increased by design. The company's image and communication is improved as well. The English design performance study by the Design Innovation Group (DIG), mentions benefits for using design such as "reduced manufacturing costs, increased profit margins, and improved external image" (Potter, 1992) and (Roy, 1994).

Figure 1 synthesises the design literature. The model shows the contributions of design in different stages of the product's life cycle. It illustrates the contribution of design both to the company and to the user.

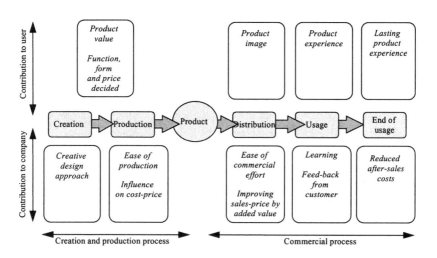

Figure 1 Design's influence and contribution towards company and user

In the product creation process, design contributes with new ideas for products (Lorentz, 1990). The "creative design approach" means that design inspires the organisation by working as a catalyst and by forming links among disciplines (Lorentz, 1990). Design helps to push and move business forwards (Logan, 1997). A Dutch design efficiency study found that design "speeds up" the development process and brings "rigour" to the process (Ginkel, 1997). Even in this age of powerful workstations, the designers drawing skills facilitate communication and help the idea generation. The creative design approach includes as well the designer's synthetic skills and suspicion towards traditional solutions (Lorentz, 1990).

The influence of design on production can result in 'ease of production' and a 'lower cost-price.' This can be achieved by reducing material costs, improving production costs (Ginkel, 1997), using appropriate materials (Goodrich, 1994) and by the standardisation of components (Buijs, 1996).

During the creation and production phase the 'product's value' is determined, because these two phases determine the product's function, form and cost-price.

During distribution, design delivers advantages for the company because of 'ease of commercial effort.' A convincing product needs less sales effort, gives free publicity, motivates the sales-force (gives them self-assurance), and might even motivate others in the company. Another important aspect of the contribution of design for the company can be a 'higher sales-price,' if the design's added value and differentiation enables the product to be positioned at a higher price point (Walsh, 1988).

Design helps the user at the stage of distribution to form an 'image of the product.' Design influences the perceived product performance (Walsh, Roy and Bruce, 1988). The product appearance, its symbolic value, its perceived function, reliability, quality and safety convince the client to select the product over competing products.

During usage the 'product experience' takes place. The product's technical performance, its user-friendly conception (ergonomics and ease of use), its reliability and experienced pleasure and status will appear once the consumer starts using the product. During this phase the company can have a 'learning experience' from the consumers' reactions.

When the product approaches its end of usage, design can contribute to the user with a 'lasting product experience' with aspects such as liability (Goodrich, 1994), durability, costs of use and ease of maintenance. For the company, design can deliver 'reduced after-sales costs' that can be obtained by an intelligent conception of the product and by a reduction of the company's environmental costs (Ginkel, 1997).

Positioning industrial design between art and R & D

The French expression for industrial design is 'esthétique industrielle.' This expression focuses on the aesthetic aspect of design. The German Bauhaus had an even broader perception of design. As in 'esthétique industrielle,' the original Bauhaus products were designed for industry. However the Bauhaus philosophy went one step further and followed the concept 'form follows function.' This not only includes product aesthetics, but product functionality as well. Functionality comes even first and the styling aspects follow the product's functionality.

Figure 2 positions Industrial Design among other important creation and innovative processes. The vertical axis represents the degree to which the product's shape is dominated by functional constraints. A primarily functional product, such as a Bic ball-point, is positioned high on this vertical axis. The shape of the Bic ball-point is dominated by its basic function of writing and influenced by technical constraints to make the product inexpensive to produce. An example of an object, indifferent to functionality, and positioned at the low end of the vertical axis, is a sculpture.

The horizontal axis represents the degree to which the product's form is influenced by emotional, affective aspects. Products on the right of the horizontal axis do not leave the user indifferent. In the relation between user and product, "feeling" and emotion is involved. An example is a Mont Blanc fountain pen. The Mont Blanc fountain pen gives its user a feeling of luxury. The shape of the Mont Blanc fountain pen is influenced by good ergonomics, but symbolic elements in the product design make the product special for its user and contribute to the pleasure of writing.

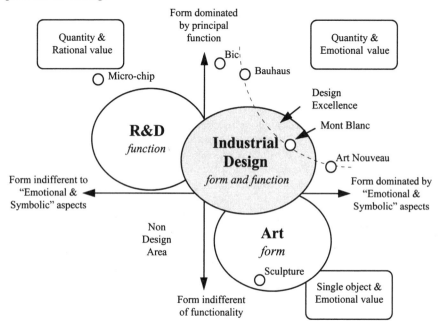

Figure 2 Three creation processes and the place of industrial design.

Research and development (R&D) is positioned where the function of the product defines the shape of the product. In R&D the form is rational and dominated by technical production constraints. A typical R&D example is the creation of a microchip. As long as this chip is not made visible or is not directly in contact with the client, the emotional appearance and shape of the product are of little importance. This product development has a purely technical orientation. If this microchip lies hidden in an electronic appliance, the specific expertise of an industrial designer it not needed.

There is an overlap between R&D and industrial design. Once the product becomes visible or in direct contact with its user the expertise of an industrial designer can be needed to improve the relation between the product and its user as product designers are trained to bring this user-orientation into the product development process.

In industrial design, both form and function are important. The shape of the product is influenced by human aspects, such as ergonomics and cultural symbolic signals. These become necessary to make the product attractive in the market place. The form, or the emotion in the product, is strongly influenced by culture, consumer expectations and company values.

An example of this balance between form and function and the degree of 'emotion' or 'functionality' are the prior mentioned writing tools, the ball-point and the fountain pen. An inexpensive Bic ball-point might write as well (or even better under certain circumstances) as an expensive Mont Blanc fountain pen. Their basic writing function is more or less the same, but their emotional value is very different.

Another example is the design of a Porsche car. The emotional aspect in the shape of the car is very important, and even its basic function (driving or fast driving) has a strong emotional aspect. The shape of the car underlines even more the car's possibilities.

Art is positioned where practically only emotion or form is important, (an artist is indifferent to the object's functionality). An example is a sculpture. Its emotional expression is important, but the sculpture is not created to be functional. Between art and industrial design an overlap exists. Both art and industrial design reflect symbols and feelings of society. However art is more free in its expressions as is not limited by technical production constraints, ergonomics or product safety. Historically the roots of industrial design are to be found in art.

Summarised; visible or direct contact with the product means the need for industrial design to improve the relation between the product and its user.

THE DESIGN INSPIRATION MODEL

This second part is used to present a model that contains the different drivers or input factors for new ideas in product design.

A model describing two different input factors for new product development, namely 'pull' and 'push' factors, has been used by Hetzel (1995). If the impulse for the development of a new product comes from a demand in the society this is called "pull." Product development mostly initiated by the company itself is called "push." Considering this idea of "push" and "pull," it is possible to propose a new model in which design literature can easily be categorised.

The key words chosen for this new model are culture, competition, users, management & human resources, technology, company values, design process and results because:

- Factors outside the company that activate the company for product innovation are *culture, competition* and *users*.
- Factors inside the company that give new impulses or influence product design are *management & human resources, technology* and *company values*.
- During the design management *process* these six factors are merged and this gives *results* to the company.

The factors inside and outside the company converge and lead to the generating of new ideas in product development. The designer translates the society's pull and company's push into a new product proposition. In figure 3, the "New Product Design" leads to company performance. Product success and company performance not only depend on design quality, other factors like the introduction strategy Hultink (1997), also heavily influence the final results. To simplify the diagram these other influencing factors are left out.

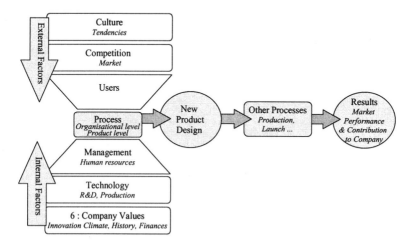

Diagram 3 Proposed model for input of new ideas in industrial design.

LITERATURE SUPPORTING THE INSPIRATION MODEL

With the "design inspiration" model (figure 3) we get an overview on the design-oriented literature. The literature is summarised using the key words of the model. Therefore this overview is divided into:
- external input factors such as *culture, competition, users,*
- internal input factors such as *management, technology, company values,*
- *design process* and *results.*

External factors / culture, competition and user

Culture, competition and users are external factors that activate the company for product innovation. The 'culture overview' includes literature on design movements, history of design and designers, and concerns the largest collection of literature.

Culture

Culture has a strong influence on industrial design. Art Deco and Streamline styling are good examples of the influence of culture on products. Both styling movements were influenced by the economic health of the society.

'Art Deco' formed a mental escape from poverty in the crisis years. This style resulted in mass-produced inexpensive products that seemed 'luxurious,' using extravagant 'Hollywood' symbols.

The 'Streamline' represented a powerful style of the United States confident in its future and technology (of cars and planes) after World War II. This style initiated by Loewy, was first applied to 'dynamic' consumer products such as cars, to be used in a later stage in 'static' products such as pencil sharpeners.

The culture section of table 1 is divided in five sections, namely "Design and Society", "Design and Country", "Styling Movements", "Designers" and "Trends".

Table 1 Design literature focusing on culture's influence on product design.

Culture				
History				Actual
Design & Society	Design & Country	Styling movements (Art)	Designers	Trends (Social & Demographic)
Woodham, 1997	Overview	Style overview	Gaudi	Green Design
Tambini, 1996	Aldersey-Williams, 1992	Sparke, 1986	Zerbst, R. 1992	Dermony, J. & Hanmer-Lloyd, S. 1997
Pile, 1994		Memphis		
Guidot, 1994	Matheu, 1993	Horn, 1986	Loewy	
Dormer, 1993	Japan	Biodesign	Jodard, P. 1992	Cramer, J. 1997
de Noblet, 1988, 1993	Dietz & Monninger, 1992	Colani, 1981	Mackintosh	Whiteley, N. 1993
Sparke, 1986, 1987		Bauhaus	Wilhilde, E. 1995	
Beaudet, 1986	Izumi, 1989	Droste, 1990		Mackenzie, D. 1991
Margolin, 1984	Raison, 1989	Westphal, 1990	Starck	
Heskett, 1980	Sparke, 1987	Art Deco	Boissière, O. 1990	Silver Design
	Former East Germany	Horsham, 1990		Buijs & Valkenburg 1996
	Bertsch, 1990	Klein et al. 1991		
	USA	Art Nouveau		Pirkl, 1991
	Sexton, 1987	Haslam, 1989		
		Perception of form		
		Smets, 1986		

About 'Design and Society': Woodham (1997), Sparke (1986, 1987) and Heskett (1980) place design movements in their context. They analyse for example the Art Deco, Streamline and Bauhaus and relate these movements to situations in the society such as welfare or crisis and to the introduction of new production technologies. Dormer (1993) completes these overviews by making links between graphic design and movements in society. Pile (1994), Tambini (1996) and Noblet (1993) describe products, designers and design-oriented companies.

As regards 'Design and Country': Aldersey-Williams (1992) makes an excellent (often cited) overview of the use of design in different countries. He illustrates the nationalistic or global aspects in product design. Often literature only concentrates on design related to one specific country such as Japan (Sparke, 1987) or former East German design (Bertsch, 1990).

The literature in the 'Styling Movements' category is close to art books. Sparke (1986) gives a clear overview of different styles in product design. Often the focus is on one specific style, par example Art Deco (Klein *et al.,* 1991) or the Italian style Memphis (Horn, 1986). Smets (1986) investigates how people perceive a product's shape.

With regards to 'Designers,' a great number of books exist. These books deliver background information about the work of a specific designer. Sometimes a designer initiates a styling movement, like Loewy. This type of historic description makes clear how pioneers in design such as Loewy (Jodard, 1992) and Mackintosh (Wilhilde, 1995) worked.

Social trends come from cultural values. Two trends spring out to have influence on product design, green ecological design and golden design (meaning design for ageing consumers).

Companies face growing pressure to make their products and services 'greener.' This pressure from society makes companies and designers improve products on ecological aspects. This not only means the design of cleaner, more energy-efficient products, but also the improvement of the psychological life-span of products and the understanding of factors contributing to the timelessness of an object. Whiteley (1993) summarises and gives examples of the use of ecological aspects in product design. Dermony and Hanmer-Lloyd (1997) give an overview of literature related to environmental responsibility within new product development. Their quantitative research concerns the integration of environmental responsibility into product development. These environmental issues in product design pose considerable difficulties for managers. Golden Design and Silver Design are terms used in Japan for products designed for ageing people. Silver stands for 70 year old and golden for 80 year old persons (Buijs and Valkenburg, 1996). Life expectancy and spending level for ageing people are growing. Redesigning and rethinking products does not only improve products for ageing consumers, but can improve products for all ages.

Competition

The competition section of the literature (Table 2) is divided into two parts, namely: 'Design-oriented companies' using design to improve their competitive strength and 'Design agencies' delivering design services.

Input for new product ideas comes from competition in two different ways. Firstly the analysis of competitors' products brings good ideas for new product designs. Secondly competition makes innovation (and design) necessary. An example is the Swiss industry that produced the same cuckoo clocks for a long time. This industry had to stand up against Asian companies, producing cheaper watches, which resulted in the design-oriented Swatch industry.

Information on 'Design-oriented companies' such as Philips (Heskett, 1989), Alessi (Gabra, 1994), Swatch (Hayak *et al.*, 1991) and Thonet (Vegesack, 1996) show the competitive advantage design gives those companies. Numerous case studies are available on the use of design in design-oriented companies. Often the focus is only on one company, as the case study on Braun (Freeze, 1992). This case study analyses Braun's way of working and its principles for good design. Jevnaker (1995) analyses the integrating of design in the furniture company Stokke. Design gains more and more its place in this company, as the innovative product designs prove to be successful. In their thesis Dubuisson and Hennion (1996) investigate the Renault design centre and two design agencies. Thackara (1997) describe the companies that participated in the 1997 European design prize. This gives a good picture of the design-oriented companies in Europe.

The 'Design Market' in England is analysed by Bruce and Morris (1996). Slappendel (1996) investigates this situation in the New Zealand design market. The Federation of French Design contributes with an overview of design in France (Laborne, 1994).

Table 2 Design literature focusing on the use of design in competition

Competition			
Design-oriented companies			Design agencies
Focus on one company (books)	One company case-studies (articles)	Focus on several companies	
Alessi,	Babolet	Thackara, 1997	Design market
Gabra ea, 1994	Hetzel & Wissler, 1997	Dubuisson & Hennion, 1996	Bruce, & Morris, 1996
Philips,			Slappendel, 1996
Heskett, J. 1989	Thomson	Buijs & Valkenburg, 1996	Design agencies
Swatch	Logan, 1997	Pearce, C. 1991	Caron, 1992
Hayak, et al. 1991	Stokke		International competition
Thonet	Jevnaker, 1995		Nussbaum, 1997
von Vegesack, 1996	Northern Telecom		Gorb, 1995
	March, 1994		
	Braun		
	Freeze, 1992		
	British Rails		
	Potter, Walsh, 1991		

Users

The product development process changes from having the company and designer central to having the user central (Buijs and Valkenburg, 1996). Like an actor, a designer tries to put himself in the shoes of the potential user. Literature about users and design are not extensively included in this table since this concerns marketing-oriented literature. Only literature is selected that describe specific methods to facilitate the integration of the user-orientation in the product development process.

Leonard and Rayport (1997) describe 'observation techniques' to integrate the user in the product development. Observing customers that use products (by preference in their own environment), helps to understand what users need. This brings new ideas for product development. An example; ' observers saw people combining beepers and cell phones not to answer calls but to screen them'.

Revat (1997) describes what could be called the 'story-telling method.' This method is used by Salomon Ski to imagine the target-group for product development. In a story the type of skier and his habits are told and this helps the development team to identify with this person in order to understand what this person likes in the product.

Dumas (1994) recommends a 'totem-building' technique. Pictures of objects in a certain style are selected to visualise the design direction. This method helps the design team to achieve focus and to understand the project goals at the early stages of product development.

Consumer-driven Japanese product design is illustrated by Takeuchi (1991). Snelders (1995) analyses the subjectivity in the consumer's judgement of products.

Internal factors / management, technology and company values

Three levels of internal factors (within the company) are distinguished that have influence on the development of new product ideas: "management and human resources", "technological developments" and "company's cultural values" (Table 3).

Table 3 Design literature focusing on the internal factors involved in product development

Internal factors		
Management & Human Resources	Technological Developments	Company Culture & Values
Management and Design Dickson Schneider, Lawrence, Hytry, 1995 Bailetti & Guild, 1991 Dumas & Whitfield, 1989 Hart & Service, 1988 Teamwork Logan, 1997 March, 1991	Buijs & Valkenburg, 1996 Manzini, 1989	Sutton & Kelley, 1997 Hertenstein & Platt, 1997 Dano, 1996 Hetzel, 1995 Hetzel & Marion, 1993 Semprini, 1992 Floch, 1990

Management and human resources

'Management and design' concerns design-oriented literature investigating the attitude towards design by management or the status of design in the organisation. Dickson *et al.,* (1995) found that 7 out of 10 general managers involved in manufacturing believe that design issues will be of increasing importance for US firms' competitiveness in the coming decades. However only 22 % felt that their managers had adequate training to make design decisions. Close to half indicated that they believe that design should be incorporated into the MBA curriculum.

Dumas and Whitfield (1989) sought to investigate current practice and attitudes towards the management of design in British industry. They found four different design approaches. The existence of a design manager has serious influence upon attitudes towards design in the company. Hart and Service (1988) focused on the attitudes and the values of top management towards product design and found seven basic managerial orientations. The "balanced orientation" (technology-push and market-pull) shows to be the most successful orientation.

Other research in this field concerns 'teamwork.' March (1991) focuses on the individual energy and talent of individuals in a team. As "some teams work better than others," we can learn about effective team management, shared responsibility and how to nurture commitment from different individuals in a team. However mutual respect and trust remain the real underpinnings of any successful team. Logan (1997) explains that most products are a "complex balance of ergonomics, aesthetics, and technology." His company has build an organisation that complements its existing structure "with an eclectic mix of humanists, artists, and technologists," as no single professional skill is sufficient to design successful products.

Technological developments

Technological developments give opportunities for new product ideas and change the profession of an industrial designer. Technological developments to speed up the development process are virtual product development teams and, 24-hours development team (using video-conferencing and e-mail). Rapid prototyping techniques give opportunities as well for the quick production of design models. Market-driven technology that influences product design are customisation and flexible production. Products can be more clever and user-friendly by using micro-electronics (Buijs and Valkenburg, 1996).

Company culture & values

Sutton and Kelley (1997), describe the culture and creative methods used in the very successful American design agency IDEO. Designers instead of doing creative work in isolation, work while clients, reporters, students and researchers visit the office. Normally working under audience increases anxiety and decreases creativity. At IDEO it works in the opposite way. IDEO has a specific company creative culture. Some companies even immerse their own designers in the IDEO culture for six weeks to improve their product development skills. Semprini (1992) translates company values in a strategy for communication which can be used for product design as well.

Design process

Concerning the design process in design-oriented companies, two levels are distinguished: the "organisational level" , and the "product level" (Table 4).

Table 4 Design literature focusing on process

Design Process		
Organisational level		Product level
Design Management (books)	Design Management (articles)	Function & Form
Design Management	**Research**	**Function**
Bruce & Cooper, 1997	Trueman, 1998	March, 1994
Cooper & Press, 1995	Brun, 1998	**Form**
Bruce & Biemans, 1995	Cooper & Press, 1997	Veryzer, 1997
Chaptal de Chanteloup, 1993	Cegarra & Hetzel, 1997	Loosschilder, Schoormans, 1995
Walsh, Roy, Bruce, Potter, 1992	Walsh, 1995, 1996	Babayou & Beaudouin,
Oakley, 1990	Lorentz, 1994	Collerie de Borely & Renault,
Borja de Mozota, 1990	Langrish, 1992	Stepanczak & Volantier, 1995
Vitrac & Gaté, 1989	Gorb, 1990, 1991	Beckwith, 1994
Lorentz, 1990	Endt, 1990	Dumas, 1994
Quarante, 1984	Walsh, Roy & Bruce, 1988	Luh, 1994
Creativity Process	Gorb & Dumas, 1987	Crozier, 1994
Kao, 1997	Kotler, Rath, 1984	Sanders, 1992
Roozenburg & Eekels 1991	Dumas & Mintzberg	Berkowith, 1987
Rapport	**Journal**	
Brun, Cegarra, Hetzel & Wissler, 1997	Beaufils, 1996	
	Grandadam, 1993	

Organisational level

Among the French books concerning design management, Chaptal de Chanteloup (1993), Borja de Mozota (1990), Vitrac and Gaté (1989), and Quarante (1984) deliver general accounts of design's competitive contribution and describe different issues in design management. Brun, Cegarra, Hetzel and Wissler (1997) analyse the design development process.

The English contribution, concerning issues and methods in design management, comes from Bruce and Cooper (1997), Cooper and Press (1995), Bruce and Biemans (1995), Oakley (1990), Lorentz (1990), and Walsh, Roy, Bruce and Potter (1992). Roozenburg and Eekels (1991) analyse design methodology and the design process.

Among the articles concerning process at organisational level, one of the first articles was written by Kotler and Rath (1984). They propose a methodology for a introduces the term silent design. This means design tasks are performed by a variety of people next to their normal work.

Dumas and Mintzberg consider the "Infusion" level the best way of managing design in an organisation. Design infusion is described as "the permeation of design throughout the organisation. Infusion is informal and everyone in the organisation is concerned by design." "Managers whose responsibilities touch design do not merely accept it, but become part of it. Design thus becomes a way of life in the organisation."

Trueman (1998) proposes in a four-level model to classify the attributes of design. She uses this model to audit companies. Walsh (1995, 1996) analyses the relationship between design and other functions in the company, and notice differences in approach between marketing and design. Walsh, Roy and Bruce (1988) analyse the contribution of design in more detail. Brun (1998) describes design management for small companies.

Product level

Literature at the product level describes the different aspects in the product such as usability, and emotional affective aspects in product design. March (1994) describes a case-study to improve product design by making the product very user-friendly and easy to understand.

Emotionality and desirability of products are analysed by Beakwith (1994) and Sanders (1992), the process of liking a product by Crozier (1994), and Luh (1994) relates the balance between function and form to the product's life-cycle. These findings concerning the emotional value of the product are particularly interesting as they add important information to the existing new product development literature.

In research on new product development Cooper (1995, 1996) defines a "differentiated product with unique customer benefits and superior value for the user" as a very important critical factor leading to excellent product performance. The design-oriented literature that analyses the desirability level in the product is therefore very important.

Sanders (1992) concentrates on three different product aspects for the product to be successful. The product should be useful, usable and desirable. In three phrase she describes her concept clearly:

- A *useful* product is one that consumers need and will use.
- A *usable* product is one they can either use immediately or learn to use readily.
- A *desirable* product is one they want.

She mentions video cameras as examples. These cameras are useful and desirable, but not always easy to use. Another example are "hospital-type" bathroom taps targeted for elderly persons. These hospital-type bathroom fixtures work, but lack appeal to their user.

She links her useful, usable and desirable concept with the Maslow pyramid of needs. Maslow describes in his model different levels of needs, from basic needs such as the need for food, to more advanced needs such as comfort, belonging to a groups and the need for social status. However we could think that to be successful a product does not need to be perfect or to be useful, usable and desirable all at the same time. We all buy products, although we do not need them. We accept products that could have been designed in a better way, but we accept those products because they are desirable. Desirability makes a lower degree of usability acceptable. The Volkswagen Beetle and the Chaise Thonet (created in 1858 !) have been time-resistant over generations. But these products are far from functionally perfect. Try the ergonomics of a Thonet's chair or the fuel-consumption of the Beetle. Maybe a lower degree of usability is part of their charms and emotional appeal.

Beakwith (1994) describes the use of psychologists, anthropologists and sociologists at Philips in their product development process. By working with psychologists, anthropologists and sociologists, Marzano's design staff (Marzano is head of Philips design) clearly aims to satisfy psychological and emotional needs in products.

Crozier (1994) in his book "Manufactured Pleasures, Psychological Responses to Design" analyses emotional and psychological needs in products. He analyses the process of liking or disliking a product and how we get used to new products shapes. Why do products stay our favourite ones ? Why do we still like a car like the Volkswagen Beetle 50 years after its conception? Why do other designs change? What is the patterns of getting used to shapes that we rejected in the first place? Why do products give pleasure? Why are objects designed the way they are? Crozier tries to bring these questions back to Freud's theories on human behaviour and Berlyne's research on reactions to artwork. He relates the existing (often psychology) literature and this delivers some understanding of the process.

Looschilder and Schoormans (1995) describe two types of product consequences; functional consequences and psycho-social consequences. Functional consequences are more or less objective. Psycho-social aspects concern elements such as the perception of status, are often symbolic, and to some extent emotional and subjective.

Luh (1994) analyses the changing balance between form and function during the product's life cycle. Over time the product's role is changing in society and in every next stage of the product's life cycle the importance of form characteristics grow:

- In the introduction phase, the product performs as a *new tool* and the product's *function* makes the product wanted.
- During the growth phase the product becomes *standard equipment* and to make the product desirable an appealing product shape is needed as well. In this stage we need a *balance between form and function.*
- The maturity phase means, that the product becomes a status-reflector and form becomes more important than functionality.
- At the decline phase the product becomes an *entertainment source.* The product is desirable because of its *form* characteristics even if the product functions badly or not at all.

Luh (1994) uses is the Barbie doll as an example. First Barbies performed as a new tool, in a traditional dolls market, allowing functions as hair combing. As Barbies became popular, even girls who did not really liked them felt the pressure to have one (standard equipment for small girls). In the third stage Barbie has "status accessories" such as a house and a car. In the future if another kind of doll or toy becomes more popular, Barbie can become purely a collector's item (entertainment source) which is no longer used for its basic function of playing.

The company Samsung and the designer Starck use the affective elements in product design. Samsung invested in emotional or psychological expressive aesthetics by working with ZIBA Design, an American top agency. One result of their co-operation is an unconventional television chosen as one of the "World's Best Products" by the Industrial Design Society of America (Nussbaum, 1997). Starck in "Mettre de l'amour dans le produit" explains that his role as product designer is more in semiology than in aesthetics (Bommel, 1997).

Improving the affective aspect in the product value is far from the only factor leading to product success. Buijs and Valkenburg (1996) mention an increasing number of factors and an increasing number of actors involved in product development. Functionality and facility in production are important as are other factors such safety and comfort, longer lasting recyclable products and the possibility of   product repairs. The increasing number of actors in product development often includes external partners such as suppliers, clients, and distributors.

Design results

Finally several design performance studies, such as the French study, "Les PMI françaises et le Design" by (Cote-Colisson *et al.,* 1995) and "The Benefits and Costs of Investment in Design" by (Potter *et al.,* 1991) measure the "results" of design activities (table 5).

Table 5 Design literature focusing on results

Results		
Govermental or Design Organisation	DIG Design Innovation Group	Thesis or article
Crouwel, Ginkel, 1997 Cote-Colisson & Le Louche, 1995 Goodrich, 1994	Roy, 1994 Roy & Potter, 1993 Potter & Roy & Capon & Bruce & Walsh & Lewis, 1991 Potter, 1992	Ainamo, 1996 Yamahoto & Lambert, 1994

FURTHER RESEARCH

The design literature informs us first, that companies are having "difficulties managing design" (Lorentz, 1994), (Dickson, Schneier, Lawrence and Hytry, 1995), and (Dumas and Mintzberg). In the second place we learn, that "product superiority" is very important as critical factor for excellent product performance (Cooper, 1995, 1996). In the third place we learn, that the inclusion "of emotional symbolic aspects" contributes to the value of the product (Sanders, 1992), (Beakwith, 1994), (Crozier, 1994) and (Luh, 1994).

Difficulties in managing design

Companies are having difficulties integrating design, even after design has proved its commercial power since 1930. Today's design-conscious companies have moved through several ever deeper stages of design management (Lorentz, 1994).

Dumas and Mintzberg consider the infusion of design in its organisation as the most effective way of using design in a company. As an example, they present the functioning of a fine restaurant in France. Everyone in this restaurant could be considered as 'designer.' The decor, the menu, the courses, the table settings are all designed. The placement of the food on the plate is not just decided in the kitchen. The waiters carefully arrange the cheeses they have just cut. In this restaurant, "design" is everyone's responsibility and a matter of care and attitude that defines the quality of the restaurant. "Design is more an attitude in the company that permeates all activities" (Dumas and Mintzberg).

Dickson et al. (1995) found that 70% of general managers involved in manufacturing believe that design issues will be of increasing importance for US firm's competitiveness, but felt that their managers had adequate training to make design decisions.

We could think of products having a 'hard' functional side and a 'soft' affective side. With growing world-wide competition, this emotional 'soft' side is important for product differentiation. Companies are having less problems in managing the rational hard side of products, as management is often trained in 'hard' skills, such as rational, verbal, and analytical skills. The management of the 'soft' side in the product demands visual, synthetic and psycho-socio skills that carry a lower social status (Lorentz, 1994).

A superior product is a critical success factor

In the new product development literature, Cooper (1995) defines the critical factors leading to excellent product performance. A "high quality new product process", a "superior product" and "true cross-functional team" were found to be very important success factors. The importance of a superior product was found in Cooper's earlier research as well. Cooper (1996) mentions "rarely are steps built in the development process that encourage the design of a superior product".

Emotional, symbolic aspects contribute to the product's value

In the design-oriented literature, Sanders (1992), Beakwith (1994), Crozier (1994) and Luh (1994) analyse the role of form and function in the product. Their findings concerning product aesthetics and the emotional value of the product add important information to the existing new product development literature. Improving the emotional aspects to the product contributes to the product's value.

CONCLUSION

We hypothesise that steps to obtain a 'superior product' are rarely built into the product development process because management is 'not comfortable with design aspects', such as the inclusion of 'emotional aspects in the product.'

Therefore the management of the inclusion of emotional factors into the design process deserves more research attention.

Research should look at: How to manage the important, subjective, aesthetic and emotional aspects in the product? How to get the balance between functionality and emotionality in the product right? How can managers more effectively incorporate the emotional and symbolic elements into the design process? How do design-oriented companies act to incorporate these subjective aspects?

REFERENCES

Ainamo, A., 1996, *Industrial Design and Business Performance*, (Helsinki School of Economics and Business Administration).

Aldersey-Williams, H., 1992, *World Design: Nationalism and Globalism in Design,* (New York: Edition Rizzoli).

Babayou, P., Beauouin, V., Collerie de Borely, A., Renault, C., Stepanczak, C. and Volatier, J.L., 1995, *Design et forme naturelle de l'objet: la mise au point d'un outil de design et d'ingénierie de l'immatériel*, (Paris: Credoc).

Bailetti, A. and Guild, P., 1991, Designer's Impressions of Direct Contact Between Product Designers and Champions of Innovation. In *Journal of Product Innovation Management*, No. 8, pp. 91-103.

Baudet, H., 1986, *Een vertrouwde wereld: 100 jaar innovatie in Nederland*, (Amsterdam: Uitgeverij Bert Bakker).

Beaufils, P., 1996, Le design dope la compétitivité. In *Industries et techniques*, No. 767, pp. 56-59.

Beckwith, D., 1994, Putting the Hard Edge on Soft values: The Higher Order of Cross-Functional Multidisciplinary Teams. In *Design Management Journal*, fall, pp. 10-16.

Berkowitz, M., 1987, Product Shape as Design Innovation Strategy. In *Journal of Product Innovation Management*, No 4, pp. 274-283.

Bertsch, G., Hedler, E. and Dietz, M., 1990, SED: Schönes Einheits Design, *Stunning Eastern Design, Savoir Eviter le Design,* (Köln: Taschen,).

Boissiere, O., 1990, *Philippe Starck*, (Köln: Taschen,).

Bommel, S., 1997, Philippe Starck: Mettre de l'amour dans le produit. In *L'essentiel de Management,* pp. 122-125.

Borja de Mozota, B., 1990, *Design & Management,* (Paris: Les Editions d'Organisation).

Bruce, M. and Biemans, W.G., 1995, *Product Development; Meeting the Challenge of the Design-Marketing Interface,* (Chichester: John Wiley & Sons Ltd).

Bruce, M. and Cooper, R., 1997, *Marketing and Design Management,* (London: Thomson).

Bruce, M. and Morris, B., 1996, Challenges and Trends Facing the UK Design Profession. In *Technology Analysis & Strategic Management,* No. 4, pp. 407 - 423.

Brun, M., 1998, Design Management: les PME aussi. In *Revue Française de Gestion,* January-February, pp. 30-42.

Brun, M., Cegarra, J.J., Hetzel, P. and Wissler, M., 1997, *Conception de Produit-Design, Rapport de synthèse,* Ministère de la Recherche.

Buijs, J. and Valkenburg, R. 1996, *Integrale Productontwikkeling,* (Utrecht: Uitgeverij Lemma).

Caron, G., 1992, *Un Carré Noir dans le Design,* (Paris: Dunod).

Cegarra, J.J. and Hetzel, P., 1997, La Gestion des projets de design: déroulement des projets et modalités de fonctionnement. In *Gestion 2000,* September-October, pp. 109-131.

Chaptal de Chanteloup, C., 1993, *Le stratégie du profit et du plaisir,* (Paris: Dunod).

Colani, L., 1981, For a Brighter Tomorrow. In *Car Styling*, No 34.

Cooper, R. and Press, M:, 1995, *The Design Agenda, A Guide to Successful Design Management,* (Chichester: John Wiley & Sons Ltd).

Cooper, R. and Press, M., 1997, Design as Strategic Resource for Management. In *Gestion 2000*, September-October, pp.91-108.

Cooper, R.G. and Kleinschmidt E., 1995, Benchmarking the Firm's Critical Success Factors in new Product Development. In *Journal of Product Innovation Management*, No. 12, pp. 374-391.

Cooper, R.G. and Kleinschmidt E., 1995, Performance Typologies of New Product Projects. In *Industrial Marketing Management*, No 24, pp.439-456.

Cooper, R.G. and Kleinschmidt E., 1996, Winning Businesses in Product Development: The Critical Success Factors. In *Research Technology Management*, No. 4, pp. 18-29.

Cooper, R.G., 1996, Overhauling the New Product Process. In *Industrial Marketing Management*, No. 6, pp. 465-482.

Cote-Colisson, D. and Le Louche, A., 1995, *Les PMI françaises et le design*. (Paris: Ministère de l'industrie,).

Cramer, J., 1997, Stretch-methode voor 'groene' producten. In *De Ingenieur*, No. 8, pp. 34 - 35.

Crozier, R., 1994, *Manufactured Pleasures, Psychological Responses to Design*, (Manchester and New York: Manchester University Press).

Crouwel, W., 1997, Industriele ontwerpers vragen meer erkenning. In *NRC Handelsblad*, 9 June.

Dano, F., 1996, Packaging: une approche sémiotique. In *Recherche et Applications en Marketing*, No.1, pp. 23 - 35.

Dermody, J. and Hanmer-Lloyd, S., 1995, Developing Environmentally Responsible New Products: The Challenge for the 1990s. In Bruce, M. and Biemans, W.G., 1995, *Product Development; Meeting the Challenge of the Design-Marketing Interface*, (Chichester: John Wiley & Sons). pp. 289-323.

Dickson, P., Schneier, W., Lawrence, P. and Hytry, R., 1995, Managing Design in Small High-Growth Companies. In *Journal of Product Innovation Management*, No. 12, pp.406-414.

Dietz, M., Monninger, M., 1990, *Japan Design*, (Köln: Taschen,)

Dormer, P., 1993, *Design Since 1945*, (London: Thames and Hudson).

Droste, M., 1990, *Bauhaus 1919 - 1933*, (Köln: Taschen,).

Dubuisson, S. and Hennion, A., 1996, *Le design: l'objet dans l'usage*, (Paris: Ecole des Mines).

Dumas, A. and Mintzberg, H., (unknown), Managing Design; Designing Management. In *Design Management Journal*, pp. 37-43.

Dumas, A. and Whitfield, A., 1989, Why design is Difficult to Manage: A Survey of Attitudes and Practices in British Industry. In *European Management Journal*, Vol. 7, No. 1, pp. 50-56.

Dumas, A., 1994, Building Totems: Metaphor-Making in Product Development. In *Design Management Journal*, Winter, pp. 71-82.

Endt, E., 1990, Design: du «cosmetique» au stratégique. In *Revue française de Gestion*, No. 80, pp. 94-100.

Floch, J. M., 1990, *Semiotique marketing et communication: sous les signes, les strategies*, (Paris: Presses Universitaires de France).

Freeze, K., Through the Backdoor: The Strategic Power of Case Studies in Design Management. In *Design Management Journal*, Fall, pp. 26-34.

Gabra-Liddell, M., Bettella, A. and Hodges, N., 1996, *Alessi, ontwerpers, design en productie*. Van Buuren Uitgeverij, Weert. Original title *Alessi, the fun factory*, (London: The Academy Group).

Ginketl, D., 1997, Industrieel ontwerp brengt MKB economisch succes. In *Design in Business*, pp.25-27

Goodrich, K., 1994, The Design of the Decade: Quantifying Design Impact Over Ten Years. In *Design Management Journal*, Spring: pp. 47-55.

Gorb, P., 1990, Design-management et gestion des organisations. In *Revue française de Gestion*, No 80, pp. 67-72.

Gorb, P., 1991, Le "design-management". In *Problèmes économiques*, No. 2.209, pp. 25-29.

Gorb, P., 1995, Managing Design in an Uncertain World. In *European Management Journal*, No.1, pp.120 - 127.

Gorb, P., and Dumas, A., 1987, Silent Design. In Bruce, M. and Cooper, R., 1997, *Marketing and Design Management*, (London: Thomson), pp.159-174.

Grandadam, S., 1993, Le design, une fonction mal assise dans l'entreprise, In *Les Echos*, mardi 28 septembre, p. 37-38.

Grange, T. 1996, Le signe, l'art et le design de l'automobile, In *Les cahiers du Management Technologique*, No. 17, p. 47-55.

Guidot, R., 1994, *Histoire du Design 1940-1990*, (Paris: Edition Hazan).

Hart, S. and Service, L., 1988, The Effects of Managerial Attitudes to Design on Company Performance. In *Journal of Marketing Management*, **4**, No 2, pp. 217-229.

Haslam, M., 1989, *Art Nouveau: Ontstaan, ontwikkeling en opleving*, (Atrium, Alphen aan den Rijn).

Hayak, N., Calabrese, O. and Schifferli, C., 1991, *Swatch after Swatch after Swatch*, (Milan:Electra).

Hertenstein, J. and Platt, M., 1997, Developing a Strategic Design Culture, In *Design Management Journal*, spring, pp. 10-19.

Heskett, J., 1989, *Philips, a Study of the Corporate Management of Design*, (London: Trefoil Publications).

Heskett, J., 1980, *Industrial Design*, (London: Thames and Hudson).

Hetzel, P. and Wissler, M., 1997, Rôle et intégration du design extérieur. In *Gestion 2000*, September-October, pp.149-164.

Hetzel, P., 1995, Pour renouveler les processus d'innovation en entreprise. In *Revue française de Gestion*, April-May, pp. 87 – 98.

Horn, R., 1986, *Memphis, Objects, Furniture and Patterns*, (New York: Simon and Schuster).

Horsham, M., 1990, *Le style des annees 20 et 30*, (Courbevoie: Editions Soline).

Hultink, E.J., 1997, *Launch Strategies and New Product Performance*, Thesis University of Technology Delft.

Izumi, S., 1989, *Packaging Design in Japan*, (Köln: Taschen).

Jevnaker, B., 1995, Developing Capabilities For Innovative Product Designs: A Case Study of the Scandinavian Furniture Industry. In Bruce, M. and Biemans, W.G., 1995, *Product Development; Meeting the Challenge of the Design-Marketing Interface*, (Chichester: John Wiley & Sons Ltd), pp. 181-201.

Jodard, P., 1992, *Raymond Loewy: Most Advanced Yet Acceptable*, (London: Trefoil Publications).

Kao, J., 1997, *Jamming: The Art and Discipline of Business Creativity*, (London: Harper Collins).

Klein, D., McClelland, N. and Haslam, M., 1991, *Art Deco, Ontstaan, ontwikkeling en opleving van deze decoratieve stijl*, (Atrium, Alphen aan den Rijn). Original title *In the Deco Style*, (London: Thames and Hudson).

Kotler, P., Rath, A., 1984, Design: a Powerful but Neglected Strategic Tool. In Bruce, M. and Cooper, R., 1997, *Marketing and Design Management*, (London: Thomson), pp. 204-214.

Laborne, A., 1994, *Livre Blanc des professionnels du Design Français: Proposition pour une politique de développement du Design Français*, (Paris: Fédération Française du Design).

Langrish, J., 1992, Design Management-Synthesis Squared. In *Design Management Journal*, Fall, pp. 10-13.

Leonard, D., and Rayport J., 1997, Spark Innovation Through Empathic Design. In *Harvard Business Review*, November-December, pp. 102-113.

Logan, R., 1997, Research, Design and Business Strategy. In *Design Management Journal*, Spring, pp. 34-39.

Loosschilder, G. and Schoormans, J., 1995, A Means-End Chain Approach to Concept Testing. In Bruce, M. and Biemans, W.G., 1995, *Product Development; Meeting the Challenge of the Design-Marketing Interface*, (Chichester: John Wiley & Sons Ltd), pp. 117-132.

Lorentz, C., 1994, Harnessing Design as Strategic Resource. In *Longe Range Planning*, Vol. 27, No. 5, pp. 73-84.

Lorenz, C., 1990, *La dimension design*, (Paris: Les editions d'organisation), (original title, The Design Dimension).

Luh, D.B., 1994, The Development of Psychological Indexes for Product Design and the Concepts for Product Phases. In *Design Management Journal*, Winter, pp. 30-39.

Mackenzie, D., 1991, *Green Design: Design for the Environment*, (London: Laurence King).

Manzini, E., 1989, *La matière de l'invention*, (Paris: Centre Georges Pompidou).

March, A., 1991, Some Teams Work Better Than Others. In *Design Management Journal*, Spring, pp. 49-53.

March, A., 1994, Paradoxical Leadership, a journey with John Tyson. In *Design Management Journal*, fall, pp. 17-22.

March, A., 1994, Usability: The New Dimension of Product Design. In *Harvard Business Review*, September-October, pp.144 – 149.

Margolin, V., 1984, *Design Discourse*, (Chicago and London: The University of Chicago Press).

Matheu, M., 1993, Design et Entreprise. In *Réalités Industrielles*, No. 1.

Noblet de , J., 1993, *Design, Miroir du Siècle*, (Paris: Flammarion).

Noblet de, J., 1988, *Design: Le geste et le compas*, (Paris:Editions Aimery Somogy).

Nussbaum, B., 1997, Design, The World's Best Products. In *Business Week, European Edition*, pp. 38-52.

Oakley, M., 1990, Design Management, a handbook of issues and methods, (Oxford: Basil Blackwell).

Pearce, C., 1991, Design classics van de twintigste eeuw, (Atrium, Alphen aan den Rijn). Original title, Twentieth Century Design Classics, (London: Blossom).

Peters T., 1994, *The Pursuit of WOW !*, (New York: Vintage Original).

Pile, J., 1994, *The Dictionaiy of 20th Century Design*, (New York: Da Capo Press).

Pirkl, J., 1991, Transgenerational Design: A Design Strategy Whose Time Has Arrived. In *Design Management Journal*, Fall, pp. 55-60.

Potter S. and Walsh, V., 1991, Design Teams on the Rails. In *Design Management Journal*, Spring, pp. 24-28.

Potter, S., 1992, Using a Sample Survey in Design Management Research. In *Design Management Journal*, Fall, pp. 35-39.

Potter, S., Roy, R., Capon C., Bruce M., Walsh, V., Lewis J., 1991, *The Benefits and Costs of Investment in design*, Design Innovation Group, (Milton Keynes/Manchester: Open University and UMIST).

Quarante, D., 1984, *Eléments de design industriel*, (Paris: Maloine).

Raison, B., 1989, L'empire des objets, (Paris: Editions du May).

Revat, R., (1997), La place du consommateur dans le développement des nouveaux produits: le cas Salomon. In *Gestion 2000*, September-October, pp. 133-148.

Roozenburg, N. and Eekels, J., 1991, *Productontwerpen, structuur en methoden*, (Utrecht: Uitgeverij Lemma).

Roy, R. and Potter, S., 1993, The Commercial Impacts of Investment in Design. In Bruce, M. and Cooper, R., 1997, *Marketing and Design Management*, (London: Thomson).

Roy, R., 1994, Can the Benefits of Good Design be Quantified? In *Design Management Journal*, Spring, pp. 9-17.

Sanders, E., 1992, Converging Perspectives: Product Development Research for the 1990s. In *Design Management Journal*, fall, pp. 49-54.

Semprini, A., 1992, *Le marketing de la marque: approche sémiotique*, (Paris: Editions Liaisons).

Sexton, R., 1987, *American Style*, (San Francisco: Chronicle Books).

Slappendel, C., 1996, Industrial Design Utilization in New Zealand Firms. In *Design Studies*, No. 17, pp.3-18.

Smets, G., 1986, *Vormleer: de paradox van vorm*, (Amsterdam: Uitgeverij Bert Bakker).

Snelders, H., 1995, *Subjectivity in the Consumer's Judgement of Products*. Unpublished Thesis University of Technology Delft.

Sparke, P., 1986, *An Introduction to Design and Culture in the Twentieth Century*, (London: Unwin Hyman Ltd).

Sparke, P., 1986, *Design Source Book*, (London and Sydney: Macdonald.& Co).

Sparke, P., 1987, *Design in Context*, (London: Bloomsbury Publishing).

Sparke, P., 1987, *Japanese Design*, (London: Swallow Publishing).

Sutton, R. and Kelley, T., 1997, Creativity Doesn't Require Isolation: Why Product Designers Bring Visitors "Backstage". In *California Management Review*, Vol. 40, No. 1, fall, pp. 75-91.

Takeuchi, H., 1991, Small and Better: The Consumer-Driven Advantage in Japanese Product Design. In *Design Management Journal*, winter, pp. 62-69.

Tambini, M., 1996, *The Look of the Century*, (London: Edition Dorling Kindersley).

Thackara, J., 1997, *Winners! How Today's Successful Companies Innovate by Design*, (Amsterdam: BIS Uitgeverij).

Trueman, M., 1998, Managing Innovation by Design - how a new design typology may facilitate the product development process in industrial companies and provide a competitive advantage. In *European Journal of Innovation Management*, No. 1, pp. 44-56.

Vegesack von, A., 1996, *Thonet meubelontwerpen in gebogen hout en buisframe*. (Weert: Uitgeverij: Van Buuren). Original title Thonet. (London: Hazar Publishing).

Veyezer, R. 1997, Measuring Consumer Perceptions in the Product Development Process. In *Design Management Journal*, Spring, pp. 66-71.

Vitrac, J. P. and Gate, J.C., 1989, *Design, la stratégie produit*, (Paris: Editions Eyrolles).

Walsh, V., 1995, The relationship between Design and Innovation, *Proceeding from the COST A3 and COST A4 Workshop Lyon*, France, pp. 19-39.

Walsh, V., 1996, Design, Innovation and the Bounderies of the Firm. In *Research Policy*, No. 25, pp. 509-529.

Walsh, V., Roy, R. and Bruce, M., 1988, Competitive by Design. In *Journal of Marketing Management*, **4,** No. 2, pp. 201-216.

Walsh, V., Roy, R., Bruce M. and Potter, S., 1992, *Winning by Design: Technology, Product Design and International Competitiveness,* (Oxford: Basil Blackwell).

Westphal, U., 1990, *Het Bauhaus,* (Amsterdam: Atrium).

Whiteley, N., 1993, *Design for Society,* (London: Reaktion Books).

Wilhilde, E., 1995, *The Mackintosh Style,* (London: Pavilion Books).

Woodham, J., 1997, *Twentieth-Century Design,* (Oxford: University Press, Oxford).

Yamamota, M. and Lambert, D., 1994, The Impact of Product Aesthetics on the Evaluation of Industrial Products. In *Journal of Product Innovation Management*, No. 11, pp. 309-324.

Zerbst, R., 1992, *Gaudi-une vie en architecture*, (Köln Taschen).

Enhancing the In-House Design Capability of Industry through TCS Projects

T.G. Inns and D. Hands

The value of design as a mechanism for improving industrial competitiveness has been clearly established. Ensuring industry actually changes its behaviour, however, to take on board the design message is an unending process.

One proven mechanism for facilitating changes in the behaviour of industry is the Teaching Company Scheme (TCS). Since 1975 this part-industry, part-government funded scheme has enabled businesses to form partnerships with academia to facilitate the transfer of technology and the spread of technical and management skills. To date over 2000 TCS partnerships have been initiated, a number of which have focused on developing the design capability of the industrial partner.

This paper reports on twelve 'design related' TCS programmes which have acted as case studies in a Design Council sponsored project investigating in-house design enhancement. Each of the programmes is shown to have effected different facets of the industrial partners' business. In each study in-house design enhancement has bought clear changes to the physical attributes of the product or service offered by the business. In addition, however, through exposure to a design resource many internal changes have taken place within the companies concerned such as: Development of product development strategy; Clearer understanding of the product development process; Greater commitment to product development activity and changes to company culture and organisation.

INTRODUCTION

The value of design in improving industrial competitiveness has been demonstrated many times. Surveys conducted across different industrial sectors demonstrate the increases in turnover and profitability which can result from design investment, (Roy and Potter, 1993). Likewise at an individual company level case studies explain how individual products have benefited from consideration of design issues, (Thackera, 1997). Very often, however, for a company this message can be frustrating, despite the evidence in favour of a design approach it can be difficult to know where to start, how to get hold of design and how to manage it. The generally accepted solution to this problem is the use of an external design consultant, although often successful, evidence suggests that there can be problems with this method of injecting design into an organisation.

In 1997 the Design Research Centre, Brunel University and the School of Design Research at the University of Central England were awarded 'Special Design Project Funding' by the Design Council to investigate an alternative approach, in-house design enhancement. The project set out to examine how one particular mechanism the TCS had achieved this. Since 1975 this part-industry, part-government funded programme has enabled businesses to form partnerships with centres of academic excellence to transfer technology and management skills, (Britton and Monniot, 1991). To date over 2000 TCS partnerships have been formed covering all facets of technology and business optimisation (TCD, 1997). Amongst these projects were a number which had specifically involved transferring design skills and technologies into the industrial partner. These projects became the focus for the Design Council funded study.

PROJECT OBJECTIVES

The Design Council funded study had several objectives:
- To identify a dozen 'design related' TCS programmes which demonstrated the impact in-house design enhancement could have on products and services.
- To examine what effect in-house design enhancement had had on the company itself in each of the programmes.
- To prepare a 20 page booklet describing each of the programmes. (Inns and Hands, 1998).
- To disseminate this information through mail shots and events to industrialists, design advisors and academics. (Inns, 1998).
- To develop a web site describing each of the programmes.

This paper reports on the first two objectives. Copies of the 20 page booklet describing each of the programmes and information on the dissemination activities are available from Dr Tom Inns at the address shown above. The web site can be accessed at: www.dmc-uce.co.uk/tcd.

CASE STUDY SELECTION

Existing information on 'design related' Teaching Company Schemes was collated and categorised into themes depending on the type of design activity undertaken in the programme. In doing this use was made of the database held by the Teaching Company Directorate. Twelve TCS programmes were then chosen for inclusion in the study, these are detailed in Table 1. In making this selection the following attributes of each programme were considered:
- It was desirable that the twelve TCS programmes should cover as many different aspects of in-house design enhancement as possible. The 12 selected programmes each represent a different theme. At Montford Instruments for example the principle objective has been: 'Using design to consolidate a product range', at Zellweger Analytics Ltd., the project has focused on 'design as part of the multi-disciplinary team'.

- The TCS programmes cover many industrial sectors, programmes have therefore been chosen to demonstrate the role design can play in a broad spectrum of products ranging from automotive accessories through to traditional men's and women's clothing.
- The chosen TCS programmes also represent a wide geographical spread with examples taken from England, Scotland and Northern Ireland.
- Likewise the selected programmes demonstrate the effect of in-house design enhancement on companies ranging in size from 18 to 1600 employees.
- Finally, and perhaps most importantly the selected TCS partnerships represent projects that have been recognised as being successful and which the industrial partner was willing to have presented in case study format.

Table 1 Summary of the case studies that were included in the study.

Company	Location	No. Staff	Product	Academic Partner	Case study research
Montford Instruments Ltd.	London	18	Climatic test chambers	Brunel University	DRC
TP Activity Toys Ltd.	Stourport (Worcs)	90	Outdoor Activity Toys	Brunel University	DRC
Multifabs Survival Ltd.	Aberdeen	75	Survival suits	Brunel University	DRC
Armstrong Technology Associates Ltd	Newcastle-upon-Tyne	30	Marine Engineering Consultants	University of Newcastle	DRC
Edwin Trisk Systems Ltd.	Sunderland	110	Infra-red paint curing equipment	University of Newcastle & University of Northumbria	DRC
DAKS-Simpson Ltd.	Glasgow	1600	Traditional clothing	Strathclyde University	DRC
Trent Bathrooms Ltd	Stoke on Trent	300	Ceramic sanitaryware	Staffordshire University	DRC & UCE
Zellweger Analytics Ltd.	Poole	400	Gas detection equipment	UCE	UCE
Haley Sharpe Associates	Leicester	22	Interior Design Consultants	UCE	UCE
ASG (Accessories) Ltd.	Derby	63	Automotive Accessories	UCE	UCE
Art Glass (NI) Ltd.	Londonderry	18	Stained glass products	University of Ulster	UCE
Thorowgood Ltd.	Walsall	65	Horse saddles	Manchester Metropolitan university	UCE

RESEARCH METHODOLOGY

Information was gathered for each of the TCS programmes. This involved the following information sources:
- Background information in the form of company histories and brochures.
- Documented information on each programme was gathered from the project participants. This typically included; TCS programme proposals, minutes from Local Management Committee meetings, reports generated by the Teaching Company Associates during the programme and, where available, final reports for the programme.
- Visits were then made to the industrial partner and the academic partner during which the following questions were asked:
  - Describe how contact was initiated between industrial partner and academic partner.
  - Describe the products and services offered before the TCS programme.
  - Describe the packages of work undertaken during the programme.
  - Describe changes to products and services.
  - What tangible metric captures the changes to products and services?
  - Describe changes to the company with regard to product development: strategy, organisation, culture, commitment and process.
  - Describe key lessons that have been learnt through the TCS programme.

IMPACT OF IN-HOUSE DESIGN ENHANCEMENT ON PRODUCTS AND SERVICES

In each of the TCS programmes examined in-house design enhancement has had a significant effect on the products and services offered by the company. In some of the programmes this was very apparent with tangible changes to the company's products. At Trent Bathrooms, for example a series of new products have been developed to create an entire new product range. In others the effect has been more subtle, at Armstrong Technology Associates for example, the design service offered by the company has been enhanced with new analysis skills.

When companies were asked to identify a tangible measure of this enhancement many metrics were used to describe improvement. At some of the companies it was possible to translate the benefits of design into a direct increase in profitability, at TP Activity Toys for example profitability rose by 11%. Other companies could identify turnover increases resulting from the TCS programme, at Art Glass (NI) Ltd. for example turnover improvements of 20% were recorded. Many of the companies found it hard to use such direct measures of improvement, some of them considered such data to be commercially sensitive, others were not yet in a position to rationalise improvements into hard financial terms. Alternative measures included: Reduction in production material consumption of 12% (Thorowgood Ltd.); Reduction in number of product variants needed to fulfil customers' needs from 13 to 3, (Edwin Trisk Systems Ltd.); Reduction in assembly and component costs by 40%, (Montford Instruments Ltd.), and reduction in the number of prototypes needed during development of 10%, (DAKS Simpson Ltd.). A summary of some of the tangible improvements is given in Table 2.

Table 2 Summary of tangible improvements to products and services at each company.

Company	Tangible benefit
Montford Instruments Ltd.	40% reduction in key component costs
Zellweger Analytics Ltd.	New products for domestic markets
Trent Bathrooms Ltd.	50% increase in turnover projected within 5 years
Armstrong Technology Associates Ltd.	New in-house expertise reducing cost of previously hired in external expertise.
TP Activity Toys Ltd.	11% increase in profitability
Art Glass (NI) Ltd.	20% increase in company turnover
Haley Sharpe Associates	Increase in design efficiency (time to complete jobs)
Multifabs Survival Ltd.	Family of new products
Thorowgood Ltd.	13% reduction in material wastage in production
ASG (Accessories) Ltd.	New products for new customers
Edwin Trisk Systems Ltd.	13 products rationalised into 3
DAKS-Simpson Ltd.	10% decrease in number of prototypes

IMPACT OF IN-HOUSE DESIGN ENHANCEMENT ON THE COMPANY

In addition to improving products and services in-house design enhancement has had a direct effect on many other facets of company activity. During interviews with project participants many anecdotal references were made to how the company had changed.

In order to put these into a framework project participants were specifically asked what changes had occurred with regard to product development; strategy, culture, organisation, commitment and process as a result of in-house design enhancement. These five headings were chosen, because earlier work (Cooper and Kleinschmidt, 1995) identified these factors as key company level drivers of product development success.

In the twelve TCS programmes included in the study the following changes were observed with regard to these headings.

Effect of in-house design enhancement on strategy

Many researchers have demonstrated that the company with future vision is more likely to succeed than that focused solely on day-to-day activity. By enhancing their in-house design skills each of the companies examined in the study has developed a clearer understanding of what the future might hold. In some of the projects this was very noticeable. At Multifabs Survival Ltd., for example, the TCS programme was focused on using design to identify and visualise future trends, a significant contribution to product development strategy has consequently been made. In other programmes the presence of internal design skills has had an equally important but more subtle effect. At ASG (Accessories) Ltd., for example, in-house design has allowed new strategic alliances with customers.

Effect of in-house design enhancement on process

The importance of up-front planning is well understood. One frequently quoted statistic suggests that 85% of all future product costs be locked in place by the time a design project reaches the concept stage (Design Council, 1996). Achieving this, however, requires an understanding of the stages a project is likely to move through in its journey from idea through to product. In all of the case studies increasing the in-house design capability has made this process more apparent. At DAKS-Simpson Ltd. and TP Activity Toys Ltd. this was a fundamental part of the TCS programme, Figure 1 shows in diagrammatic form the design process that was implemented within TP Activity Toys Ltd. during the TCS programme. Even where this was less focused, however, *process* has become part of the company vocabulary as a result of in-house design exposure.

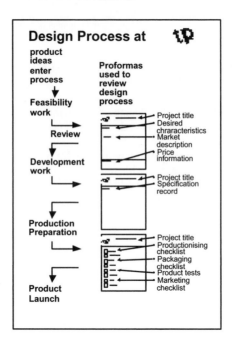

Figure 1 The three stage design process adopted by TP Activity Toys Ltd. has reduced the time
to market of new toys because more effort is now put into getting design right first time
rather than relying on time consuming iteration.

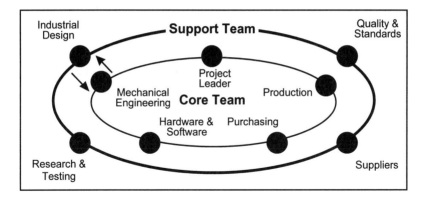

Figure 2 Product development at Zellweger Analytics Ltd. is now undertaken with active inputs from support specialists including industrial designers.

Effect of in-house design enhancement on organisation

The in-house designer has an important part to play in optimising many organisational issues, a role identified in previous studies, (Olson, 1994). At Armstrong Technology Associates Ltd. development of software based tools has broken down the technical language barriers that existed between engineering specialists. At Zellweger Analytics Ltd. the industrial design skills of visualisation have acted as a catalyst for discussion in the new multi-disciplinary project teams used by the company, this input is shown as a diagram in Figure 3. In the small company environment an injection of in-house design acts in a different way providing co-ordination across many company operations as demonstrated in the Thorowgood case study.

The effect of in-house design enhancement on product development culture

Creativity and exchange of ideas are crucial when building a culture of innovation within a company. In all the TCS programmes brining design in-house has helped revitalise these aspects of company activity. At Trent Bathrooms the product design approach has challenged traditional working practices with new ideas. The process mapping at DAKS-Simpson Ltd. has created a meeting culture encouraging people to talk freely and debate many aspects of company operation for the first time. Haley Sharpe Associates have seen the adoption of an IT system bring a more customer focused approach to design activity.

The effect of in-house design enhancement on commitment to product development

Even with strategy, process organisation and culture in place success is ultimately dependent on commitment from senior company staff. For this reason demonstration of company commitment is an important criteria in the selection of all new TCS proposals. In the 12 case studies there was, therefore, a firm commitment to the partnership objectives right from the start. In most of the TCS programmes examined this commitment was retained right the way through the programme and beyond. An indication of this is given by Table 3 that summarises the employment position of TCAs at the end of their two year TCS contracts (as of April 1998). In 75% (6 out of 8) of the completed schemes the company had committed themselves to a full-time in-house designer where before the scheme there was no one fulfilling that job description. In 100% (4 out of 4) of the schemes still in progress the company had expressed a keen desire to retain the TCA at the end of their two year TCS contract.

Table 3 Employment position of TCAs in the twelve case studies (at April 1998)

10 **OUT** **OF** **12**	**TCS still in progress (as at April 1998) - Company keen to retain TCA** - Montford Instruments Ltd. - Armstrong Technology Associates Ltd. - Edwin Trisk Systems Ltd. - DAKS-Simpson Ltd. **TCS completed (as at April 1998) - design function provided by exTCA** - Zellweger Analytics Ltd. - TP Activity Toys Ltd. - Multifabs Survival Ltd. **TCS completed - TCA left but new designer recruited as replacement** - ASG (Accessories) Ltd. - Thorowgood Ltd. - Trent Bathrooms **TCS completed - TCA left no direct replacement of skills** - Art Glass (NI) Ltd. - Haley Sharpe Associates

KEY LESSONS FROM THE CASE STUDIES

All project participants were asked to identify two key lessons which they would most like to share with others interested in design related TCS projects. A summary of the lessons from each company is shown in Table 4. What is most surprising about these statements is their diversity.

REVIEW & CONCLUSIONS

The study demonstrates that (within the sample group) the TCS programme has been an effective mechanism for enhancing in-house design. All the participants managed to identify some form of tangible benefit resulting from in-house design enhancement, usually expressed as an improvement to product(s) or service(s). Further discussion revealed considerable changes in many aspects of company operation as well. With improvements to product development: Strategy, Process, Organisation, Culture & Commitment.

From a research perspective the case studies raise a number of important issues:

- Firstly, the TCS mechanism has demonstrated its potential for allowing researchers to examine in-house design enhancement in a structured way. Each of the programmes, although addressing different industries and technologies, has been established according to the TCS regulations. This means that each company has had a similar two year burst of design activity fully reported through the meetings and reports required by the TCD. All ideal material for the researcher wishing to collate case study material.
- Secondly, The tangible benefits are expressed in a wide variety of ways by the project participants. Each of the companies was asked to identify its own metric for measuring the tangible benefits of in-house design enhancement. Due to confidentiality problems and/or the fact that direct profit and turnover increases were yet to filter through a wide variety of other measures were suggested. The message for the researcher is that a degree of creativity is often needed to extract such sensitive material.
- Thirdly, the impacts on other facets of the company were very marked, with changes to; strategy, process, organisation, culture and commitment.
- Finally the studies show the wide variety of approaches that can be taken in running 'design related' TCS programmes. The academic expertise being pumped into each programme was very different.

Table 4 Key lessons from the 12 case studies as described by project participants

Company	Key lessons (Identified by project participants)
Montford Instruments Ltd.	<ul><li>When designing new products talk to customers and challenge assumptions about their product needs</li><li>Up-front planning is needed if product potential is to be unlocked through design</li></ul>
Zellweger Analytics Ltd.	<ul><li>Undertaking product development projects with a multi-disciplinary team overcomes problems of communication and co-ordination</li><li>Building industrial design skills into the in-house team allows more effective consideration of manufacturing issues</li></ul>

Table 4 Key lessons from the 12 case studies as described by project participants (continued)

Company	Key lessons (identified by project participants)
Trent Bathrooms Ltd.	• Respect existing ways of tackling product development where these work well • The launch of a new product range requires input from all disciplines within an organisation
Armstrong Technology Associates Ltd.	• Engineering theory can be developed into workable 'design tools'. These allow rapid analysis of complex design problems • When applied at the concept stage such rapid analysis allows identification of more competitive design solutions
TP Activity Toys Ltd.	• Implementing procedures for controlling the design process can speed time to market • Considerable opportunities for improved competitiveness can be realised by systematic redesign of existing products
Art Glass (NI) Ltd.	• Design activity in a crafts-based industry can be enhanced through IT with no loss of creativity • User 'ownership' is required if implementation of an IT system is to be cost effective
Haley Sharpe Associates	• Appropriate IT systems can provide clear visualisations of design solutions to which alterations can be quickly made. This allows the customer real-time participation in the design process
Multifabs Survival Ltd.	• Successful products require careful alignment of 'technology push' and 'market pull' • Future product innovations lie on a spectrum of innovation
Thorowgood Ltd.	• Production efficiency can be increased significantly by taking product design back to first principles • In the small company environment the skills of the designer can bring leverage to many activities
ASG (Accessories) Ltd.	• By offering a design service suppliers can build stronger partnerships with their customers • Use of 2D and 3D modelling can improve communication throughout the design process. This allows more effective assessment of project proposals and early consideration of alternative manufacturing techniques
Edwin Trisk Systems Ltd.	• Seeing through an ambitious programme of product development needs full management commitment • Before embarking on a programme of product development thought must be put into defining exactly what is required
DAKS-Simpson Ltd.	• To manage and improve the design process it must be visible • Many facets of production quality can be enhanced by effective monitoring of the design process

REFERENCES

Britton, D., and Monniot, J., 1991, *The SERC/DTI Teaching Company Scheme,* Oxford, Institute of Ceramics Convention St Catherine's College.

Cooper, R.G., and Kleinschmidt, E.J., 1995, Benchmarking the firm's critical success factors in new product development. In *Journal of Product Innovation Management*, Vol. 12, pp. 374-391.

Design Council, 1996, *Business in Britain* (London: Design Council).

Inns, T.G. and Hands, D., *Design in-house: 12 case studies describing how the TCS (Teaching Company Scheme) has improved in-house design capability,* (Brunel University, Design Research Centre Publications).

Inns, T.G., 1998, Design in-house: Overview of 12 TCS case studies. In *Using TCS to enhance design in small and medium sized companies.* A one day seminar organised by TCD and sponsored by the Design Council, 8th April 1998.

Olson, E.M., 1994, Interdependence, Conflict and Conflict Resolution: Design's Relationship with R&D, Marketing and Manufacturing, *Design Management Journal,* Vol. 5, No. 4, Fall 1994.

Roy, R., and Potter, S., 1993, The commercial impacts of investment in design. In *Design Studies,* Vol. 14, **2.,** Edited by Jones, A., (Oxford: Butterworth-Heinemann), pp. 171-193

TCD 1996, *TCS Annual Report 1995/1996,* (Faringdon, Teaching Company Directorate).

Thackera, J., 1997, *Winners! How today's successful companies innovate by design,* (Aldershot: Gower).

A Key Characteristic in Co-development Performance Measurement Systems: Transparency

A.S. Johnson and S.Evans

The need to effectively manage joint product development, referred to as co-development, between vehicle manufacturers and its suppliers is widely seen in the automotive industry as a means of ensuring business survival through reduced time to market, increased quality and reduced costs. However, effectively measuring the product development process has proved difficult within industry and academia. This is attributable in part to the absence of publicly available information, and in part to the inherent difficulty of measuring a complex process lasting 3 years or more. This is in contrast to manufacturing where significant amounts of information is available and the process is less complex.

Measurement of *co-development* between a customer and supplier therefore requires a step forward in the understanding and application of measurement in the design and development process. The ability to monitor and continuously improve co-development performance is reliant on having an appropriate customer-supplier measurement system.

The author participated in the design and implementation of a Co-development Performance Measurement System. This identified the general principles of performance measurement which are important in Co-development and, more significantly, how these principles should be applied. The transparency of the Performance Measurement System was identified as one of the key characteristics.

This paper will describe the implementation of a co-development Performance Measurement System and focuses on the transparency characteristic. It will argue that when measuring co-development effectively the following are key:

- Transparency should be incorporated early in the system design
- The design/layout of the user interface must support transparency
- Transparency should be considered from an holistic stand point of system effectiveness.

and will discuss how these can be applied in practice.

INTRODUCTION

The requirement for the UK first tier automotive component suppliers to become world class at developing products with their customers (Co-development) is now seen as an order winning criteria (Foresight, 1995). Co-development is defined as "the ability of vehicle manufacturers and their suppliers to develop competitive products together" (EPSRC Grant No GR/K78553). The automotive suppliers development capability will in part determine how effective they are at competing in the global market (Davies, 1994). The automotive industry is now moving towards a structure based on partnership where profit is shared amongst the different players in the value chain. This is in contrast to the traditional structure based on confrontation (Lamming, 1986). Vehicle manufacturers (VM's) will focus their activities on the design of the car, the manufacturing of very few key components based on core competencies (e.g., engines, gear boxes, bodywork and painting) and the assembly of systems supplied by highly qualified suppliers that participate actively in the vehicle design (The Boston Consulting Group, 1993).

Effectively managing and measuring this evolving product development process is widely seen as a means of ensuring business survival through reduced time to market, increased quality and reduced costs (Driva, 1997). There is however, a lack of detail available on how measurement of product development should be approached (*ibid.*). This is attributable in part to the absence of publicly available information, and in part to the inherent difficulty of measuring a complex and creative process (Clark and Fujimoto, 1991). Oliver (Oliver *et al.,* 1996) states that there has been remarkably little work which has tried to systematically quantify design and development performance.

In addition to this sparse practice and theory in measuring product development, the measurement of co-development includes both customer and supplier and therefore requires a significant step forward in the understanding and application of measurement in the product development process (Wyatt *et al.,* 1997). The difficulty is high, but so is the need, with the ability to continuously improve co-development performance reliant on having an appropriate customer-supplier measurement system in place.

PERFORMANCE MEASUREMENT SYSTEM (PMS)

In its simplest form a PMS can be described as a closed-loop system that provides information which controls the future functioning of a process (Nalder, 1980). The fundamental elements of a closed-loop system are the process being controlled, feedback, inputs and outputs (Fig. 1). The important element in PMS terms is the feedback (Otley, 1987; Maskell, 1991; Globerson, 1985; Hanna and Burns, 1997) which can have a positive effect on performance (Pritchard *et al.,* 1988). In addition its effect on an individual's behaviour has been recognised as essential for learning and for motivation in performance-oriented organisations (Ilgen *et al.,* 1979).

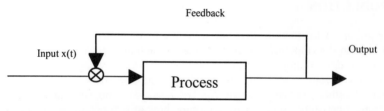

Figure 1 Closed-loop system

Maskell (1991) describes a PMS as having seven common characteristics: they are directly related to strategy; they mainly use non-financial measures; they vary between location; they change over time; they are simple and easy to use; they supply fast feedback; and they are improvement focused. The authors have previously described a PMS as having a set of twenty-two characteristics (Johnson and Evans, 1997). Three characteristics appeared regularly throughout PMS literature: the most relevant measure should be used; measures should be derived from strategy; and the effect on peoples' behaviour should be incorporated. These three characteristics may not be the most important ones but they should be considered key characteristics.

The authors have participated in the design and implementation of a co-development PMS and observed the system in operation. This took place in a vehicle manufacturer (VM) who was undergoing a major change initiative requiring an improvement in the co-development performance from its supply-base. Twenty-two characteristics of a "good" PMS for co-development were built from the literature and used as a form of checklist in the design of the measurement system (*ibid.*). In parallel with their involvement in the activity the authors conducted 54 interviews with relevant engineers, managers and directors from the VM and 9 of its suppliers. The transparency of the PMS was identified as one of the key characteristics.

TRANSPARENCY

This section will describe what transparency is and why transparency is important in a PMS.

What is it?

If something is transparent it can be seen through (*The Oxford Minidictionary*). A good example being a piece of glass. If the piece of glass is clear and free from 'bits' full vision through the glass is possible, it is said to be transparent. The piece of glass can be frosted to restrict vision but allow light to pass through. This effect is described as translucence. Taken to the extreme if the piece of glass is darkened to the extent that no light passes through it is not transparent it becomes opaque. In their work on 'Lean Thinking' Womack and Jones (1996) describe transparency as a key principle. An example they cite is that of a visual management system clearly displaying performance to target that anyone can see. This was in complete contrast to the previous system that was secretive and tightly guarded by management.

Transparency can also be described as something that is easy to understand (*The Oxford Minidictionary*). That is to say that the meaning or importance of something is seen / known / realised, or that it is clear how a system works or operates. An example being transparency of financial trading information between auction and dealer markets where the price information process is visible to all users which results in a fairer system and more importantly allows overall process improvements to be made to the system (Pagano and Roell, 1992).

This paper takes transparency as being made up of two dimensions: something that can be seen through and something that is easy to understand, and from this a model of transparency can be constructed, Fig. 2. Assuming that the purpose of using the model is to help understand how to increase the transparency of a PMS, some important points can be drawn. If the 'ease of understanding' of a PMS is increased independently of 'ease to see through' e.g., from 'Now' to point A, the system has become easy to understand but it can not be seen. An example being if a PMS has been redesigned to make it easy understand and use more effectively but the feedback to the person(s) who need to change is limited to a simple numerical score. The weakness of this approach is that although effort has been expended to increase understanding the recipient can not identify where or why a change in a score occurred only that there was a change. If an increase in the 'ease to see through' the PMS is made in isolation to the 'ease of understanding' the system, e.g., from 'Now' to point B, the system can be clearly seen but is still difficult to understand. An example being feedback from a PMS that is presented as a percentage score and has been redesigned to show the 20 variable equations with which the score is calculated.

Why have a transparent PMS?

The context in which organisations approach performance measurement systems is fundamental to the effectiveness of the system. The level of transparency of a PMS should be determined by the philosophy an organisation pursues to ensure their long-term economic well being. For example, a philosophy of continual improvement which is central to World Class Manufacturing must be supported by a PMS that encourages improvement (Maskell, 1991; Dixon *et al.,* 1992; Lynch and Cross, 1992; Stainer and Nixon, 1997). Therefore, the design of a performance measurement system must be rooted in an understanding of exactly what the organisation needs to do to exploit its sources of competitive advantage. This understanding should be communicated throughout the organisation, via the PMS, so that all employees are aware of the company goals towards which they are working (Fitzgerald and Moon, 1996). A PMS should be used to increase the level of understanding of key performance criteria throughout an organisation. Targets and the rate of process improvement should be set and communicated via the PMS (Kaydos, 1991; Sink, 1986). If the level of performance measurement system transparency is not conducive with the organisation needs (e.g., continual improvement) the system can be said to be ineffective.

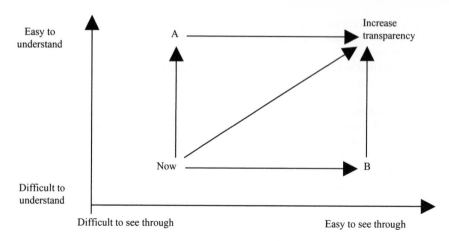

Figure 2 Transparency model

This applies not only within organisations but between organisations as in the purchaser-supplier business relationship and especially in the area of supplier development (Krause and Ellram, 1997). Effective communication is a critical aspect of successful purchaser-supplier relationships (Carter and Miller, 1989) Krause and Ellram (1997) suggest that the performance of suppliers can be significantly raised by the buying firm communicating their expectations to suppliers in an effective manner. Practitioners and theorists concur on the benefit of PMS transparency. However, achieving transparency can be challenging, especially in a product development context that bridges a customer-supplier relationship.

A CO-DEVELOPMENT PERFORMANCE MEASUREMENT SYSTEM

One global VM is effectively measuring the co-development performance of its European supply base that is in excess of 300 suppliers. The co-development performance measurement system is integrated with a supplier appraisal and improvement system that provides feedback to suppliers on their overall performance. The aim of the system is to ".. achieve corporate objectives through working with suppliers who are also seeking to establish partnerships based on mutual trust and co-operation and who have a commitment to continuous improvement following, at all times, a customer quality philosophy."

Co-development is one of five measurement criteria which a suppliers overall performance is assessed against, the other criteria being quality, cost, delivery and management. Each criterion is owned and managed by the appropriate function within the VM. The development function, for example, is responsible for assessing its suppliers' co-development performance. The overall system is co-ordinated by the purchasing department who compile quarterly feedback reports on overall supplier performance to target as well as providing specific performance feedback for each of the five measurement criteria. In addition, the system incorporates an annual Award Scheme for its highest performing suppliers.

The development function has a Supplier Development Department who are responsible for ensuring that each supplier's co-development performance is assessed accurately, consistently and reliably. The department uses a performance measurement system as a means of ensuring that the relevant data is collected, recorded, analysed and later presented to the suppliers. This is a highly structured attempt to provide clear information to suppliers, and is unusual in being applied in development.

The co-development performance measurement system has been recently redesigned. The previous system was identified as being out of line with the future strategic requirements of the VM. The supply base saw the overall supplier appraisal and improvement system as one of the best in the industry but they found that the assessment of a supplier's co-development performance to be ineffective.

Through a series of vehicle manufacturer-supplier focus groups several areas of improvement were identified, namely:

- Need to increase the level of supplier understanding of the system
- Provide more comprehensive feedback about individual suppliers co-development performance
- The level of subjective measure should be reduced.

The level of understanding that suppliers have of how their co-development performance is assessed by the VM can be attributed to the level of transparency of the system. i.e. can the system be seen through and how easy is it to understand? The need to resolve this issue led to the realisation that the VM had to become more open about how its supplier co-development performance was assessed. This was felt by the suppliers to have a strong detrimental effect on their ability to clearly identify improvement activities and directly relate these improvement activities to corresponding changes in scores received in the feedback. Therefore a desire emerged to design a co-development PMS that could be used as a base to incorporate the issue of transparency. The authors took part in this process by seeking and evaluating relevant work from around the globe and presenting this information and ideas into the VM's discussions. A new co-development scoring system was the result of this industry-university collaboration.

THE REDESIGNED CO-DEVELOPMENT PMS

The suppliers' co-development capability is assessed on an on-going basis with feedback on their performance being given every quarter. The measurement system is made up of three measurement criteria: quality, cost and delivery of each of the main stages in the VM's development process, e.g., delivery of designs, delivery of prototypes. The VM clearly states the importance it attaches to each element of its development process by weightings which indicate to the supplier where to focus its resource and effort. The previous system only provided an overall supplier performance rating which was found to be in effective. Feedback to the suppliers is presented in the form of relevant scores and in graphical form clearly showing the present and last three quarters' performance to target. Deviation from mean co-development performance across the 300 plus suppliers is also shown.

Joint reviews are conducted between the VM and supplier where alignment targets are set by the VM with specific rate of improvements agreed with individual suppliers. The reviews are also used to monitor and improve the co-development performance of suppliers (Wyatt *et al.,* 1997). The engineers input the data directly into a PC based system providing a 'real time' assessment of each supplier's performance. The system is used to record and compile reports not only on a supplier performance but is a basis for identifying improvement actions for both suppliers and the vehicle manufacturer. This is seen as key to effective co-development.

The effective measurement of co-development is additionally enabled by assessing suppliers' performance as objectively as is possible. The system is made up from 80% quantitative-objective measures the remaining measures being subjective both of which are used in an objective framework. This was a significant change from the original system that was made up of 15% objective and 85% subjective (Johnson and Evans, 1998).

TRANSPARENCY IN THE CO-DEVELOPMENT PROCESS

The consistency, reliability and credibility of the VM's development process is fundamental for effective co-development to begin to take place. The VM's development process must be understood and consistent for alignment by the development processes of suppliers. The mutual level of understanding that the VM's engineers and their corresponding suppliers' engineers have of the why and what the co-development PMS is trying to achieve is critical to the overall effectiveness of the PMS and the co-development process itself. How the system has been designed and the way in which it is presented are fundamental. The transparency of the system greatly affects the usability of the performance information that the system holds. The level and effect of the system transparency has been carefully considered in the design and operation of the co-development performance measurement system. The effects of increasing the transparency in assessing the co-development performance of its European supply base were considered not only within the development function of the VM but from a perspective of the overall effectiveness of the supplier appraisal and improvement system.

Referring to the Transparency Model (Fig. 2) the VM took route 'A' to increase the transparency of the co-development PMS. This choice was based on a strong history of general openness and information sharing plus the suppliers own identification that understandable information was a key concern (See Johnson and Evans, 1998 for more discussion on this process). It was observed that the VM's holistic approach to how the co-development PMS was designed was key to enabling the PMS transparency to be fully realised. Very early in the system design the VM had to decide how to clearly describe what their own development process was and what and where the critical supplier inputs were to this process. This was a key stage in the system design focusing the VM to ask "can we effectively communicate this to our suppliers and internally within the organisation?" The resulting system is made up of several layers of increasing detail. At the highest layer a simple single score is given for suppliers co-development performance. This is mapped over time and compared to targets and to the supplier base mean performance. At the second level a clear 3x3 matrix enables a quick and easy understanding of where scores relate to which stage of the co-development process. At the lowest layer a highly detailed system exists where specific changes in performance can be clearly identified and their origin understood. It is at this level of the system where the change in the balance of objective and subjective measures became important allowing a predominately objective analysis of suppliers co-development performance to made.

To give an example of a highly transparent co-development PMS a measure of 'on-time prototype parts' is shown in Table 1. The system is easy to understand and easy to see through.

Table1 An example of a highly transparent co-development PMS

Measure: On-time prototype parts	**Value of total score**: 2%		
Breakdown of score			
Score	0	1	3	5
Measure of score	More than 'x' days late	'x' days late	On-time part, paperwork late	On-time part & paperwork, error-free paperwork

Initial findings from 25 interviews conducted in 9 of the suppliers gave a strong indication that those suppliers who have received not only the new PMS system but also education about the co-development PMS had an increased level of understanding of the system. Their understanding had also improved their ability to effectively use the system to identify improvements and directly relate these improvements to corresponding changes in their scores. This is in comparison to suppliers who had received minimal education about the system this affected their understanding of the system and dramatically limited their ability to identify improvements.

The education process involved a combination of formal presentations and on-the-job / face-to-face discussion between VM's and supplier's engineers. showed the suppliers that the VM was willing to divulge and share previously hidden information about how suppliers performance is assessed. This 'openness' in the relationship can be compared to that seen in the use of cost transparency in the industry which also requires the sharing of information between VMs and suppliers (Lamming, 1996). Co-development PMS transparency can be described as a derivative of cost transparency where ".. information must be shared, but that the process is two-way: .... in order to develop ways (either process or product) in which cost can be removed, quality improved etc." (Lamming, 1994).

CONCLUSIONS

Co-development is referred to as a future order winning criteria for UK first tier automotive component supplier who want to compete in the global market and has been defined as "the ability of vehicle manufacturers and their suppliers to develop competitive products together". This paper has described how a vehicle manufacture is effectively measuring the co-development performance of over 300 of its suppliers. It has been identified through the authors participation and a series of interviews that the effective measurement of co-development requires a high level of PMS transparency.

The current practice is that very few organisations effectively communicate the understanding and logic of how they measure suppliers' co-development capability and that the level of PMS transparency can dramatically affect a VMs and suppliers co-development performance. Measuring product development is a challenge to both industry and academics. The amount of time and effort required to pioneer in a "difficult" area is high and the uncertainty of whether or not the benefits are greater than the input effort may cause concern amongst practitioners. The lack of widespread working examples and of the benefits of effectively measuring the product development process also plays a part.

Previous research has identified that the level of transparency of performance information presented within and between organisation is critical to achieving targets. This research has established that when effectively measuring a supplier's co-development performance a PMS should have a high level of transparency. The authors have shown that a transparent PMS will be both easy to understand and easy to see through. These simple but important concepts have been illustrated through examples from a co-development PMS now being used by 350 companies and which following its introduction, has assisted in an eight-fold increase in the rate of improvement.

REFERENCES

Carter, J.R. and Miller, J.G., 1989, The Impact of Alternative Vendor/Buyer Communication Structures on the Quality of Purchased Materials. In *Journal of the Decision Sciences,* **20**, pp. 759-776.

Clark, K.B. and Fujimoto, T., 1991, *Product Development Performance: Strategy, Organisation and Management in the World of Industry.* (Boston: Harvard Business School Press).

Davies, P., 1994, *Report to the Management Committee on the case for a Research Programme in the Manufacturing Sector of Road Vehicles.*

Dixon, J.R., Nanni, J.A.J., and Vollmann, T.E., 1992, *The New Performance Challenge: Measuring Operations for World-Class Competition.* (Homewood, Illinois: Business One Irwin).

Driva, H., 1997, *The Role of Performance Measurement during Product Design and Development in a Manufacturing Environment,* PhD Thesis. (Nottingham University: Department of Manufacturing Engineering and Operations).

EPSRC (IMI) Research proposal, Grant No GR/K78553.

Fitzgerald, L. and Moon, P., 1996, *Performance Measurement in Service Industries. Making it Work:* The Chartered Institute of Management Accounts.

Foresight, T., 1995, *Progress Through Partnership No 5 Transport,* (HMSO).

Globerson, S., 1985, *Performance Criteria and Incentive Systems,* (Oxford: Elsevier).

Hanna, V. and Burns, N.D., 1997, 1997, *The Behaviour Implications of Performance Measures,* presented at the 32nd MATADOR, Manchester 10th – 11th July.

Ilgen, D.R., Fisher, C.D. and Taylor, M.S., 1979, Consequences of Individual Feedback on Behaviour in Organisation. In *Journal of Applied Psychology,* **64**, pp. 349-371.

Johnson, A.S. and Evans, D.S., 1998, *The Objectivity / Subjectivity Balance in a Co-development Performance Measurement System,* to be presented at Engineering Design Conference, Brunel University, UK., 1998.

Johnson, A.S. and Evans, S., 1997, Metrics and Performance Measures, *presented at First International Conference on Responsive Manufacturing, Nottingham,* 17-18 September 1997, Nottingham, 1997.

Kaydos, W., 1991, *Measuring, Managing, and Maximising Performance*, (Oregon: Productivity Press).

Krause, D.R. and Ellram, L.M., 1997, Success Factors in Supplier Development. In *International Journal of Physical Distribution & Logistics Management,* **27**, pp.39-52.

Lamming, R., 1986, For Better or for Worse: Technical Change and Buyer-supplier Relationships. In *International Journal of Operations & Production Management* **6**, pp.20-29.

Lamming, R., 1994, *A Review of the Relationships between Vehicle Manufacturers and Suppliers,* (Department of Trade and Industry).

Lamming, R., 1996, Squaring Lean Supply with Supply Chain Management. In *International Journal of Operations & Production Management,* **16**, pp.183-196.

Lynch, R.L. and Cross, K.F., 1992, *Measure Up! Yard Sticks for Continuous Improvement*, (Blackwell Business).

Maskell, B.H., 1991, *Performance Measurement for World Class Manufacturing: A Model for American Companies* (Productivity Press).

Nalder, D.A., 1980, *Feedback and Organizational Development: Using Data-based Methods:* (Addison-Wesley Publishing Company).

Neely, A., Gregory, M., and Platts, K., 1995, Performance Measurement System Design. In *International Journal of Operations & Production Management,* **15**, pp.80-116.

Oliver, N., Gardiner, G., and Mills, J., 1996, Design and Development Benchmarking, presented at *Proceedings of the 3rd EurOMA Conference, 3 – 4 June,* London Business School.

Otley, D., 1987, *Accounting Control and Organisational Behaviour,* (Heinemann).

Pagano, M. and Roell, A., 1992, *Transparency and Liquidity: A Comparison of Action and Dealer Markets with Informed Trading ,* (Discussion Paper), London School of Economics.

Pritchard, R.D., Jones, S.D., Roth, P.L., Stuebing, K.K., Ekeberg, S.E., 1988, Effects of Group, Goal Setting, and Incentives on Organizational Productivity. In *Journal of Applied Psychology,* **73**, pp.337-358.

Sink, D.S., 1986, Performance and Productivity Measurement: The Art of Developing Creative Scoreboards. In *Industrial Engineering,* pp. 86-90.

Stainer, A. and Nixon, B., 1997, Productivity and Performance Measurement in R&D. In *International Journal of Technology Management,* **13**, pp. 486-496.

The Boston Consulting Group, 1993, *The Evolving Competitive Challenge for the European Automotive Components Industry: Executive Summary.*

Womack, J.P. and Jones, D.T., 1996, *Lean Thinking: Banish Waste and Create Wealth in your Corporation,* (Simon & Schuster).

Wyatt, C.M., Evans, S., Johnson, A.S., Jukes, S.A. and Foxley, K., 1997, *Co-development: The COGENT Initiative,* presented at Automotive Manufacturing Autotech '97, NEC Birmingham, 1997.

Improving Product Development Performance: two approaches to aid successful implementation

Sarah A Jukes, Fiona E Lettice & Stephen Evans

There is an increasing pressure for companies to improve their product development capability; the challenge is to provide innovative design solutions in shorter timescales whilst not compromising on product quality. Many companies are looking for a step change in their development performance, targeting improvements of around 20 - 30%. This in itself is a significant challenge, with the added pressure of insufficient assistance to achieve such improvements and limited understanding of how best practice companies achieve industry leading processes and products. This paper describes two approaches that have been developed to help companies achieve step change in product development performance. Both approaches have been developed through empirical research into the factors that affect implementation success; the paper expands on four of the factors, highlighting their significance to improvement activity and presenting their embodiment in the implementation methodologies. The selected method for guiding companies through the transition is a series of focused, interactive workshops, customised by participating companies to meet the more specific requirements of their unique business situation.

The major difference between the two approaches originates in the overall aims and objectives that each sets out to achieve. Fast CE is an implementation methodology aimed to help companies move towards multi-disciplined product development teams and to identify the Concurrent Engineering (CE) tools and techniques most suited to that company's unique environment. In contrast, COGENT is an implementation methodology developed to help an Original Equipment Manufacturer (OEM) jointly develop products more closely with its first tier suppliers through improved communication and alignment of the two companies' development practices. This approach has been termed 'co-development' and assumes that internally the collaborating organisations have relatively best practice product development processes (employing the Concurrent Engineering tools and techniques included in the Fast CE methodology) and are searching for new ways to enhance their performance.

Fast CE has been applied in 5 companies whilst COGENT has been applied in one automotive company in conjunction with almost 25% of its first-tier suppliers. The paper concludes by revealing some of the quantifiable and qualitative benefits these organisations have gained from following a structured, planned approach to product development improvement, and summarises the widespread lessons that can be learned.

INTRODUCTION

Over the last decade, a new competitive environment has emerged characterised by intense global competition, diverse markets, powerful end-customers, and rapidly changing technologies (Clark and Fujimoto 1991). Corresponding to these pressures, product development has been recognised by many organisations as a means for achieving sustainable competitive advantage, both now and in the future, and is being placed at the top of the management agenda. The challenge is to develop higher quality products in less time at minimum cost, with clear focus being placed on meeting or exceeding customer desires and expectations (Smith and Reinertsen, 1991; Adachi *et al.*, 1994).

In the hope of satisfying these demands, organisations are looking to transform their current working practices to adopt a world-class approach to product development (Susman, 1992). For many, this requires achievement of a step change in performance with targets equating to 20 - 30% improvement in product development time and cost. This is no easy accomplishment, involving the integration of each business process influencing the end-product design and the development of close working relationships across functional and organisational boundaries. To exaggerate the complexity of implementation, insufficient help exists for organisations trying to improve their product development performance, often leaving them struggling with an unstructured, unplanned implementation programme typically resulting in failure (Nichols, 1994).

This paper addresses the field of product development improvement within Western European manufacturers and will concentrate on the outcome of two research studies carried out over a six year period. The paper will initially present two methodologies which have been rigorously developed and implemented and are aimed at guiding companies through the transformation process in a structured, pragmatic way. These methodologies have been based on empirical research into the factors that influence the success of product development improvement, and a description of these factors will form the subsequent section of the paper. Finally, further insights into the process of product development improvement will be gained through presentation of the benefits that can be reaped from adopting a rationalised and planned implementation strategy.

DEFINING THE TWO IMPLEMENTATION METHODOLOGIES

Improving product development performance is becoming an ever-increasing concern of senior management in the manufacturing sector today and is dependent on how well organisations enhance the integration, collaboration, and cooperation between the different processes and disciplines involved (Lettice, 1995; Pawar, 1994). A scarcity of research exists however in the field of product development improvement, specifically to a level that could be deemed useful to potential implementors. To satisfy this current demand, two methodologies have been rigorously developed and tested that aim to secure reductions in product development lead time and cost whilst increasing the quality and sales potential of the end-product. These methodologies are outlined in the following section with differences in the intentions and circumstances surrounding each application also discussed.

The Fast CE project

The Fast CE research project commenced in August 1993 with the objective of generating an implementation methodology which guides organisations through their product development transformation towards the adoption of CE practises. This has been achieved through exploration into 'best practice' CE and the identification of low-risk, low-cost strategies that aid its rapid introduction.

What is concurrent engineering?

CE has become a widely recognised means for achieving improved product development performance by challenging the logic and practice of the traditionally sequential product development processes (Clausing, 1994). The aim is to avoid the unnatural separation of work into upstream and downstream activities through increased integration, most successfully achieved through multifunctional teamworking. CE can be defined as: "..the delivery of better, cheaper, faster products to market, by a lean way of working, using multidisciplined teams, right first time methods and parallel processing activities to consider continuously all constraints". (Evans *et al.,* 1997).

This requires radically changing the way products have been traditionally developed within western manufacturing organisations and impacts all aspects of the business.

Outline of Fast CE methodology

The Fast CE methodology (presented in Figure 1) is divided into three stages, each representing a major progression in the product development improvement process. Each stage is activated through an interactive two-day workshop which stimulates discussion, accelerates understanding, and directs concentration towards necessary action.

Stage one is the preparation stage and aims to reinforce the need to change throughout an organisations' senior management team with the intention of generating their awareness and support to the transformation process. If at the end of this phase, total commitment is not achieved, then transformation attempts should be abandoned as real business benefits are unlikely to emanate.

In the second stage, attention is drawn to the launch of the CE project team, with team members working together to understand how to create multifunctional teamworking so that effective product development can flourish. This stage is only finished when the team have completed the first pilot project and have gained experience of operating within a CE environment for every element of the development process.

Finally, the third stage of the Fast CE methodology (known as the expansion phase) captures the knowledge gained from the initial implementation and identifies how the CE principles can be carried forward into subsequent projects. The last two stages of the methodology can be repeatedly adopted to launch and review subsequent product development projects. In this way, CE is progressively introduced into the organisation, and gradually becomes the new way of working.

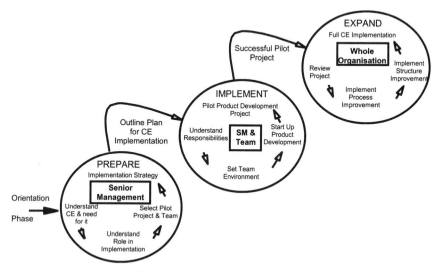

Figure 1 Fast CE implementation methodology

The COGENT project

The COGENT research project commenced in October 1995 and focused on product development improvement across organisational boundaries. COGENT is derived from Latin, meaning 'to drive forward together' and aims to rapidly improve and sustain the collaborative product development capability between an (OEM) and its first tier suppliers.

What is co-development?

Industrialists and academics alike have identified the escalating trend towards increased involvement of suppliers in the product development process (Littler and Leverick, 1993). Particularly in the automotive industry, survival and prosperity of the vehicle assemblers will depend upon whether component suppliers will take on a larger role in the extended enterprise through accepting greater responsibility for innovation and getting more deeply involved in the product development process (Leisk and Wormald, 1993). This new way of working has been termed co-development, and can be defined as: "the ability of OEMs to develop competitive products in partnership with their first-tier suppliers" (Wyatt *et al.*, 1997).

Co-development builds on the principles of CE, implying that the product development team not only involves the integration of the traditional organisational functions, but also spans across company boundaries. Through building closer customer - supplier relationships, the joint development processes can be optimised and waste eliminated, typically resulting in significant reductions to overall development time and cost. The approach assumes that co-development improvement is most suited to those organisations currently possessing capable, in-house product development practices, typically operating within a CE environment.

Outline of COGENT methodology

Similar to Fast CE, the COGENT implementation methodology incorporates the three predominant phases of the product development improvement cycle and involves the delivery of multiple, interactive workshops. The activity is lead by the sponsoring OEM who selects appropriate suppliers to participate in the improvement programme, and requests the attendance of the managing director at the first event; once an affirmative decision has been reached, COGENT inspires both parties to jointly communicate and interact at all levels within the organisation (see Figure 2).

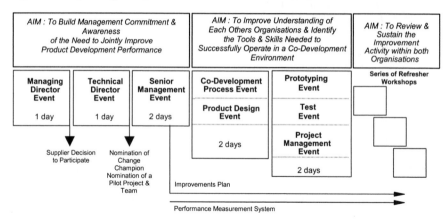

Figure 2 The COGENT implementation methodology

The initial stage of the implementation methodology aims to generate a high level of commitment towards the improvement activity within the senior management teams of the collaborating companies. Three workshops are implemented within this stage which progressively increase understanding of the future competitive environment throughout the various management tiers. At the end of this stage, joint targets and objectives for the improvement activity will have been agreed and an aligned schedule for the proposed improvement activity will have been produced including clearly defined roles and responsibilities.

Secondly, the focus of the methodology is shifted towards creating solid foundations for the future co-development relationship through the alignment of the product development processes and project management systems. Subsequently, each stage within the product development cycle is considered, with the implementation methodology providing the tools and skills required to successfully develop products collaboratively. Refinements to the improvements plan are made throughout the second stage and execution of the transformation efforts initiated.

The third phase is an iterative process which should be reused periodically based on the nature of the co-development relationship and the requirements of the organisations involved. The objective at this stage is to support the joint improvement activity through communicating successes and failures, and sustaining the transformation attempts. The emphasis shifts towards the joint measurement, monitoring and review of the original targets and milestones set in stage one with amendments to the improvements plan being aptly introduced. The co-developing partners reacquaint themselves with the concept of 'best practice' to which they are both aspiring and are encouraged to reflect on their existing strategy.

IDENTIFYING AND REALISING THE FACTORS THAT AFFECT THE SUCCESS OF PRODUCT DEVELOPMENT IMPROVEMENT ACTIVITY

Underpinning each implementation methodology is a series of factors that have been identified as influencing the success of product development improvement activity. These factors were initially generated via a thorough literature review and later validated through over 100 interviews conducted in companies worldwide, more than 50 exploratory focus groups involving industrialists occupying senior positions, and 3 months observation of actual improvement activity. Eight common factors were highlighted as significantly influencing the success of product development improvement activity; for the purpose of this paper, four have been randomly selected and expanded upon below.

Leading the change

Management commitment is often heralded as a fundamental component of transformation success and requires a significant proportion of influential people striving towards a common renewal strategy. However, if actual improvements and tangible benefits are to be realised, good intentions and ongoing discussion are simply not enough - the change needs to be put into effect and action needs to transpire. Managing and implementing any evolution process requires significant investment in time and resource, and is typically recognized as an additional activity for senior management (Crow, 1993). Efforts to transform the organisation often deteriorate as conflict arises in the day-to-day schedule and existing duties with shorter-term goals take priority over the improvement activity.

The enormity of the combined tasks instigates a need and search for assistance which is often secured through the use of change agents (Smart, 1997). The change agent should have the authority and disposition to stimulate change within the organisation, and remove potential and current barriers restricting implementation. In more successful cases, their role is primarily educational, providing support and advice, rather than getting intricately engrossed in the planning and execution of the improvement activity. Often, internal change agents are situated somewhat outside the traditional managerial hierarchy whilst maintaining strong relationships with the senior management, assisting them through the implementation of the transformation and updating them periodically on company-wide progress.

Both the Fast CE and COGENT implementation methodologies recognise the importance of securing strong leadership to oversee and manage the organisational change through use of change agents. Internal change agents, hailed as 'Change Champions', are appointed early on in the transformation process based on their organisational skills, knowledge, and status. A fundamental trait of the champion is that they have demonstrated a positive reaction to the proposed change and consider it a viable approach to best practice product development. In addition, the implementation methodologies offer guidance in defining the roles and responsibilities of the change agent and equip them with the tools and techniques which promptly activate change.

A slight anomaly arises within the COGENT framework as it requests the nomination of a change agent within both organisations engaged in the co-development relationship. Internally, the role of each champion does not deviate from the above guidelines, remaining to navigate the organisation through the change process. However, through the adoption of additional, external responsibilities, the champion's role increases in its complexity to include such tasks as the alignment of improvement activity between the co-developing partners and the eradication of cross-organisational barriers.

Pilot project

Inertia to change is often encountered when organisations are committed to the current strategy and individuals resist divergence to the status quo (Huff *et al.,* 1993). This resistance to change is often characterised by a fear of the future and a perception that the improvement activity is both time consuming and unnecessary. Significant effort is required to encourage individuals to abandon their old working practices and existing relationships.

It must be recognised that change will at first subject the organisation to inefficiencies and uncertainties; stress will increase as individuals become dissatisfied with the company's inability to respond to the external competitive pressures and the new internal environment. Single, partly isolated projects, often called pilots, can help pave the way to organisational change and overcome these significant barriers (Learnard-Barton, 1992; Kanter, 1993). Pilot projects can serve as small departures from tradition, providing a 'foundation of experience' to inspire eventual large changes (Kanter, 1983; Race and Powell, 1994). They provide an initial test for the depth of transformation possible whilst the learning that occurs within the pilot often provides the psychological safety needed by others to undergo change (Schein, 1993; Fullmer, 1993). Employment of a pilot project also means that organisational members learn through their own activities, allowing them to generate their own 'best practice' solutions which satisfy their unique problem and promote greater trust in the information discovered (Costanzo, 1993; Armenakis *et al.,* 1993).

Within the co-development environment, a pilot project needs to be agreed by both parties involved in the transformation process as it brings focus to the joint improvement activity and creates a vehicle for aligning product development processes and practices. 'Pilot project' in this context relates to a single co-development project that incorporates the product development teams within both organisations.

Both implementation methodologies support the rapid establishment and management of a pilot product development project; a single pilot success can win over remaining sceptics whilst generating a small pool of product development improvement expertise. The approach is preferred to simultaneously introducing multiple projects as it is minimises the risk and complexity of managing the improvement activity.

Action orientation

Jaffe & Scott observed that many managers attended workshops to develop change management skills and ideas, which equipped them with the enthusiasm and excitement to change, only returning to their workplaces to experience the old environment and prevailing relationships (Jaffe and Scott, 1993). This 'workshop high' tends to produce momentary resolve to the change predicament, followed by cynicism and disappointment on return to the real world setting. Off-site, intensive training has been evaluated by a number of academics as not achieving the desired changes in attitudes and behaviours, with participants finding it difficult to take what they have learned back to the workplace.

Senior managers need to take the lead in change programmes by recognising and examining how their own behaviours and actions contribute to the organisation's problems (Argyris, 1991). This can be best achieved by realising how counterproductive the consequences of defensive reasoning and external blaming mechanisms are in contrast to productive reasoning and admission of one's own shortcomings. Through constructive analysis of real product development projects and previous complications experienced, productive reasoning can be achieved; the challenge is in learning how to learn.

A requirement of both implementation methodologies was to avoid the 'workshop high' syndrome and encourage more productive reasoning. The approaches should be based on the premise that learning is intimately related to action and material presented should focus directly on the users' needs so that it is perceived as relevant to the unique business situation in which they are operating. Creating the ability to apply the learning immediately on return to the organisation means that results materialise quickly, building enthusiasm for further learning and change. This cycle of understanding the problem, action evaluation and rapid implementation is continuous and gradually changes the behaviours and culture within or between the organisations involved. It is easier to achieve a culture change in this way rather than identifying culture change as the objective of the learning process.

Fast CE and COGENT initially concentrate on the identification of positive actions that management and the pilot project team can implement to increase the likelihood of a rapid and successful launch of the transformation process. This includes the definition of roles and responsibilities within the change programme, use of structured problem solving, and assignment of clear milestones and targets for the improvement activity (including how costs and benefits will be shared).

Group participation

"Transformation is impossible unless hundreds of people are willing to help, often to the point of making short term sacrifices. Employees will not make sacrifices, even if they are unhappy with the status quo, unless they believe that change is possible" (Kotter, 1995).

One method to build commitment to the improvement activity is to establish a high level of ownership for the changes required with those directly affected by the new way of working (Backhouse and Brokes, 1996). One requirement of an effective methodology would therefore be to construct a process that builds ownership and accountability for implementation of the improvement activity. Group participation in implementation planning will increase acceptance and lead to a common understanding of the necessary action (Platts, 1990). Involvement at this early stage often leads to a transferral of ownership through to the latter phases of the transformation and has been shown to reduce the risk of implementation failure.

Implementing CE is a cross functional change process and requires that individuals are drawn from their traditional departmental roles into multi-functional product development teams. This is also a true statement when operating in a co-development environment, but an additional complexity is introduced when the transformation process requires involvement at a multi-company level. In either instance, change can not be driven from one department but requires the support of the managing director and his board of senior managers. Pilot project members should work alongside senior managers to ensure that a cross section of views on the change process is acquired and ownership for implementation is widespread. This will also help to break down some of the traditional, hierarchical and cross-functional barriers that exist both within and between organisations and will introduce a process of negotiation between the parties involved.

Due to the multi-company focus of co-development, it is important to stress that the success of the COGENT implementation methodology is not only dependent on the degree of ownership at all levels within a single organisation, but requires commitment and ownership by both parties involved in the change. Improvement activity should add benefit to both organisations and a win-win scenario needs to be established.

Workshops were selected as the main tool for instigating improvement at each stage in the implementation process as they provide a mechanism by which the above factors can be incorporated into a low cost, human-centred approach. Workshop participants, carefully selected based on their influence on the transformation efforts, are subjected to a series of presentations and group activities which promote participation and involvement. Checklists, pro forma, and ongoing documentation ensure that solutions are generated rapidly and effectively whilst facilitating transferral of the learning back into the workplace. Following each workshop, implementation of the improvement activity is dependent on the organisational setting, but is supported by further methods such as regular progress reviews, collocation of the development teams, and adoption of relevant performance measures.

APPLICATION OF THE METHODOLOGIES & EXISTING OUTCOMES

The Fast CE and COGENT methodologies have repeatedly been applied within manufacturing organisations around Western Europe with ongoing implementation based on persisting demand. The following section discusses the implementation strategies of both methodologies and presents the quantitative and qualitative results that have emerged.

Fast CE implementation & results

The Fast CE implementation methodology initially evolved through a use-refine-use-refine process and during this period was applied in four discrete manufacturing organisations operating within diverse industrial sectors. From this initial development and implementation, a workbook has been constructed which supports the introduction of CE from its initial launch through to sustainable application. To date, the workbook has been comprehensively applied within three organisations, with further exploitation proposed within the automotive industry.

To date, company A has deducted eight months from their product development lead time saving them an estimated £7 million, whilst company B realised a saving of £8 million. For most companies however it is too early for quantifiable benefits to be determined, but many have reported perceived early benefits from adopting the Fast CE methodology. These include increased motivation within the project teams, improved communication across the disciplines and a more productive atmosphere for product development to flourish. Company C regard their product and implementation plans for the pilot project superior to those generated for past development projects, and report a "positive feeling" towards the new approach. Another organisation stated that "We have achieved interim results that are very encouraging. We have got better results than we expected at this point in the programme ... it's not a gimmick but a different way of working and people really see the benefits from it".

Cogent implementation & results

To date, the development and implementation of the COGENT methodology has focused on the automotive industry, in particular it has tried to improve the co-development performance of one vehicle manufacturer and approximately 25% of its first tier suppliers. The second phase of the COGENT application strategy is currently underway and aims to apply the implementation methodology in other OEMs and their supply base to test its relevance outside the initial application.

Success of the COGENT activity is being measured and monitored through a number of mechanisms; these include utilisation of a quantitative performance measurement system held at the vehicle manufacturer and regular progress reviews which gauge improvement activity against critical targets. Causal analysis in a complex area such as co-development is inevitably a difficult task, yet the data is beginning to demonstrate a positive divergence in the rate of co-development improvement between the COGENT suppliers and the control group. The targets set for 10% reductions per annum in overall product development time and cost over a three year period are being exceeded, and a total of 74 improvement actions have already been implemented across the supply base. Perceived benefits have also been recognised, with users of the COGENT methodology conveying that it has improved the relationship between the co-developing partners and increased understanding of the joint development processes. Participants have commented that the approach has promoted 'open communication with the customer' whilst 'focusing attention on the important issues' and 'increasing management awareness of the need to improve' - all recognised as important for co-development success.

CONCLUSIONS

The research presented here has uncovered a number of factors that affect the success of product development improvement activity within and between manufacturing organisations. Two methodologies have been discussed which incorporate these influential factors, the aim to rapidly improve product development performance, in the first instance through employment of multi-functional teams and CE practices, and secondly through building closer, collaborative relationships between OEM's and their first tier suppliers. The lessons learnt are portrayed in Figure 3 and include recommendations of tools and techniques that can be adopted to effectively realise the influential factors. "Implementation, conventionally thought of as what happens after a decision has been reached concerning organisational action, is crucial to organisational change and effectiveness .. implementation differences can mean the difference between success and failure" (Sproull and Hofmeister, 1986).

Organisations that are intent on making substantial improvements in their product development processes must invest in the skills, capabilities, and mind sets that are essential to achieve efficient and effective product development (Clark, 1989). This is a complex, challenging journey, comprising of many barriers and pitfalls, but successful organisations reap the benefits of rapidly introducing distinctive products which often exceed customer expectations. The methodologies presented in this paper do not provide a simple solution, but aim to support organisations attempting to achieve this transition and guide them through the common implementation challenges.

What?	Where?		How?
Factors Affecting Success of Product Development Improvement Activity	**Concurrent Engineering Context**	**Co-Development Context**	**Useful Implementation Techniques**
1. Provide Clear, Effective Leadership to Manage the Transformation	Single Company Involvement Internal Focus to Product Development Improvement Multi-Functional Integration	Multi-Company Involvement External/Internal Focus to Improvement Activity Joint Alignment of Development Processes, Projects & Teams	Senior management to attend a workshop Nomination of Change Champion
2. Select a Pilot Project to Focus Improvement Activity			Select a real project Nominate a team leader & members Identify project aims
3. Involve all Personnel affected by the Change to Build Ownership & Commitment			Involve team in decisions Provide team with necessary tools & skills Jointly identify roles & responsibilities
4. Take an Action Oriented Approach to Increase Awareness & Understanding of the Unique Business Environment			Early development of an implementation plan Use examples & past experiences Use the development project to define targets

Figure 3 influencing the success of product development implementation

REFERENCES

Adachi , T., *et al.*, 1994, Strategy for Supporting Organisation and Structuring of Development Teams in Concurrent Engineering. In *The International Journal of Human Factors in Manufacturing.*

Argyris, C., 1991, Teaching Smart People to Learn. In *Harvard Business Review*, Vol. 69, No 3, pp. 99 – 109, May/June.

Armenakis, A. A., *et al.,* 1993, Creating Readiness for Organisational Change, *Human Relations,* Vol. 46.

Backhouse, C. J. and Brookes, N.J., 1996, *Concurrent Engineering: What's Working Where*, (Gower Publishing Ltd).

Clark, K.B. and Fujimoto, T., 1991, *Product Development Performance - Strategy, Organisation and Management in the World Auto Industry,* (Boston, Massachusetts: Harvard Business School Press).

Clark, K.B.,1989, Project Scope and Project Performance: The Effects of Parts Strategy and Supplier Involvement on Product Development, In *Management Science*, Vol. 35, No. 10, pp. 1247 – 1263.

Clausing, D. P., 1994, *Total Quality Deployment : A Step-by-Step Guide to World Class Concurrent Engineering,* (New York : ASME).

Costanzo, L., 1993, Breaking Out is Hard to do. In *Engineering*, April.

Crow, K. A., 1993, Implementing Concurrent Engineering : Lessons Learnt. In *Concurrent Engineering and CALS Conference,* Washington DC, June.

Evans, S., Lettice, F.E., & Smart P.K., 1997, *Using Concurrent Engineering for Better Product Development,* In CIM Institute Report, (Cranfield University).

Fullmer, D. M., 1993, Large Scale Implementation of Concurrent Engineering at Martin Marietta *Concurrent Engineering and CALS Conference,* 21st May.

Huff, J. O., Huff, A.S. and Thomas, H., 1992, Strategic Renewal and Interaction of Cumulative Stress and Inertia. In *Strategic Management Journal,* Vol. 13, pp. 55 – 75.

Jaffe, D. T., and Scott, C.D., 1993, Building a Committed Workplace : An Empowered Organisation as a Competitive Advantage. In *The New Paradigm in Business : Emerging Strategies for Leadership and Organisational Change,* Ray, E.M. and Rinzler, A., (New York : Pedigree Books).

Kanter, R. M., 1983, *The Change Masters,* (New York : Simon and Schuster).

Kotter, J. P., 1995, Leading Change : Why Transformation Efforts Fail. In *Harvard Business Review,* Vol. 73, No. 2, pp. 59 – 67, March/April.

Learnard-Barton, D., 1992, Core Capabilities and Core Rigidities : A Paradox in Managing New Product Development. In *Strategic Management Journal,* Vol. 13, pp. 111 – 125.

Leisk, C., and Wormald, J., 1993, *Supplier Innovation : Building Tomorrow's Components Industry,* (London: The Economist Intelligence Unit).

Lettice, F.E., 1995, *Concurrent Engineering : A Team Based Approach to Rapid Implementation,* PhD Thesis, Cranfield University.

Littler, D., and Leverick, L., 1993, Factors Affecting the Process of Collaborative Product Development. In *Journal of Product Innovation Management,* **28** (3).

Nichols, K., 1994, Developing with the Best : Survey of Product Development within UK Industry. In *World Class Design to Manufacture,* Vol 1.

Pawar, K. S., 1994, *Implementation for Concurrent Engineering in the European Context,* Department of Manufacturing Engineering and Operations Management, (University of Nottingham).

Platts, K. W., 1990, *Manufacturing Audit in the Process of Strategy Formation,* PhD. Thesis, Cambridge University.

Race, P., and Powell, J., 1994, Observations in Implementation of SE, the Use of Gaming Techniques to Empower Teams and Case Studies of Team Working, *IEE Colloquium on Issues of Cooperative Working,* June, London.

Schein, E. H., 1993, How Can Organisations Learn Faster? The Challenge of Entering the Green Room. In *Sloan Management Review,* Winter.

Smart, P. K., 1997, *An Empirical Investigation of the Factors that Contribute to a Successful Implementation of Concurrent Engineering,* PhD Thesis, Cranfield University.

Smith, P. G., and Reinertsen, D.G., 1991, *Developing Products in Half the Time,* (New York : Van Nostrand Reinhold).

Sproull, L. S., and Hofmeister, K.R., 1986, Thinking about Implementation. In *Journal of Management,* Vol **12**, No 1.

Susman, G. I., 1992, *Integrating Design and Manufacturing for Competitive Advantage,* (New York : Oxford University Press).

Wyatt, C., *et al.*, 1997, Co-Development : The COGENT Initiative. *Autotech Conference Proceedings, Automotive Manufacturing Section*, Birmingham, November.

One Phone Number for Europe: cultural diversity, technology and innovation.

John D. Law

At a recent UN forum, a leading American politician observed that negotiations with European governments were made more difficult because each nation spoke with a different voice.   In frustration, the US delegate allegedly commented, "What we need is one telephone number for Europe". This viewpoint highlights some of the divisions that remain between the 15 member states of the European Union (EU).   Five years after deregulation of the single internal market, many differences in political and economic strategies prevail, and individual countries continue to be characterised by distinct markets, consumers and cultural diversity. Historically, Europe has proved to be a rich source of technological innovation, which has been highly dependent upon the catalysts of research, design and development. The European Commission is now well established as a valid political force and is set to play a dominant role in regulatory matters.

Cecchini, (1988) estimated that a fully integrated market could realise an increase of 50% in private sector investment in science and technology being directed towards improving the competitiveness of European companies.   The European Commission recognises that some regions of the EU are in strong positions to benefit from the single market, particularly those with prominent infrastructures and established industrial bases.   This is typical of Western Germany, the Benelux countries, Eastern France, Southern UK and parts of Northern Italy. The commission's policy over the last quinquennium has sought to address this imbalance by allowing provision for grants to be made available to areas on the periphery of the EU.   Consequently, a number of funded schemes (administered through Joint Research Centres, (JRCs), in Germany, Netherlands, Italy and Belgium) now exist to promote research and development, which would normally be beyond the resources of individual member states.

This paper examines the impact of this strategy upon the ability of EU member states to manage new product innovation and investigates the link between innovation, demographics and cultural diversity.   Would the 'one telephone number' approach to research and development be beneficial in managing new product innovation throughout Europe?

INTRODUCTION

The notion that cultural diversity influences people's perceptions of product design has been well established within the design profession (Papaneck, 1971; Henley Centre, 1992). The rich variation of social, political and cultural traditions across the EU offer many different interpretations of product style, performance and quality. There have been many extraordinary transformations brought about by increased political and economic union in Western and Central Europe over the last decade. The continued expansion of the Single European Market (SEM) has presented pan-European manufacturers with an enigma in matching product identities to 15 different nations within the EU. Treaties with the remaining European Free Trade Area (EFTA) nations, now offer an extended European Economic Area (EEA) creating a potential single market of 18 countries with 370 million consumers.

Historically, Europe has been the origin of many technological innovations, which have been realised through the application of research, design and development. An industry's technological environment can be described in a variety of ways, many of which focus on the time related progression of that environment. One definition involves an evolutionary approach (Nelson and Winter, 1982), suggesting that technology within an industry develops along a 'technological trajectory' and can be described according to its position on that trajectory. This helps to explain the considerable variation in levels of technological sophistication across the EU, with individual countries evolving at different points along 'technological trajectories' that in turn help to shape opportunities for product innovation.

However, successful innovation is dependent on a much broader spectrum of issues than solely research and development. Organizational structure, concurrent approaches to production and marketing, financial resources, and design management will all impact upon the outcomes of the innovation process. Bacon and Butler (1981) argue; "..invention is the solution to a problem, often a technical one, whereas innovation is the commercially successful use of the solution."

Oakley (1990) expands on this relationship between invention, innovation and design and comments; "From the start, new product development must focus on innovation and continuous product development. By examining all aspects of innovation prior to major expenditures on development - such as market need, technical, production and marketing requirements, and a basis for protecting the product from competition - the product solution can be found."

TECNOLOGICAL INNOVATION

A definition of technological innovation, offered by the Organization for Economic Cooperation and Development (OECD, 1981), describes this activity as; ".. the transformation of an idea into a new or improved saleable product or operational process in industry or commerce."

Part 10 of BS 7000 (BSI, 1989) 'Design Management Terminology' includes the word 'novel' in a similar definition of innovation;".. transformation of an idea into a novel saleable product or operational process in industry and commerce or into a new service."

Both definitions link the concept of embodying a new idea or principle in a product with a commercially competitive context. The use of the term 'novel' implies that innovation will take longer than evolutionary design and will require a greater level of investment. The time lag between investment and return is also generally increased. Technological innovation carries a higher risk than NPD and there are many examples which demonstrate that innovation alone, cannot guarantee success in a marketplace (BR's advanced passenger train, Hutchinson's 'Rabbit' mobile phone). New developments, unless appropriately protected, can be quickly copied, limiting the extent to which an innovating manufacturer can gain extra profit accruing to a successful innovation before they are reduced through imitation. 'Technological opportunity' is defined by Davies (1993) as; "..the inherent potential for further profitable development of the industry's product technology into new applications and process technology into cost reductions or performance enhancement."

Davies links this with the maturity of the technology; new technologies having a high level of opportunity which diminishes as the scope for applications becomes fully exploited. Dominant designs arise with regard to products and the opportunities for cost reductions through production, improvements are expended. 'Innovative' products can render competitors products obsolete, through applications of new technologies or materials.

Research and development

Research and Development (R&D) is defined by the OECD as an activity which comprises; "..creative work undertaken on a systematic basis in order to increase the stock of knowledge."

The OECD link the use of such knowledge to providing new or substantially improved processes, products, systems, materials and services. Many companies invest in R&D because they expect this to lead to innovation. However, this may not necessarily be the case, since not all new knowledge is appropriate to *commercial* products or services. Companies may invest substantial resources in R&D only to discover limitations in a new product that will render it unworkable in a commercial context.

The adoption of a Concurrent Engineering approach can change the way in which participants in the NPD process relate to each other (Hughes, 1995). Concurrency encourages multidisciplinary teamwork as a replacement to the segregated sequential progress of traditional programmes. However, integrating a Concurrent Engineering approach within technological innovation is problematic since innovative aspects of a product may prove unworkable or take longer than expected to productionize. To maintain overall time schedules, design teams may be pressured into reducing time allocated to testing. This carries the associated dangers of going to market before fully testing and resolving difficulties linked to the innovative (untried) aspects of the product (Potter and Roy, 1986).

Hollins and Hollins (1995) suggest that this problem can be tackled by identifying those aspects of a new product design which are innovative, and allocating extra time to them at the beginning of the development process. Once sufficient confidence in the progress of innovation is established, the subsequent phases of this part of the development process can be run in parallel with the evolutionary aspects of the design. The European objectives set by a company will determine the profile of its strategy, focusing expensive resources on priority projects that will strengthen the company's competitive position in the SEM.

In setting an R&D strategy, a company must review its ability to utilise existing facilities, technology and in-house experience in meeting project objectives. 'Gaps' may be addressed by reviewing opportunities for acquiring new technologies appropriate to the project via in-house investment in capital equipment and training. Alternatively, collaborative ventures or license agreements may provide solutions. Corporate venturing is a relatively new concept in Europe, whereby a major company can seek to access new technology and acquire new product lines by acquiring a minority share in a smaller company.

Developed in the US over the last decade, corporate venturing is not constrained by national or EU borders. European companies can utilise venture partners outside the EU, accessing new R&D resources and creating potential for entry into foreign markets. The projected capital investment in realising each option, aligned with European objectives, must be assessed carefully. This in turn, will influence the composition of manufacturing and marketing criteria in the development of a group of strategies for the execution of new product development programmes. A company may wish to rationalise its products and ranges in respect of harmonised European standards.

MANAGING INNOVATION - A LEGISLATIVE APPROACH

Most European Commission staff operate in the 23 Directorates-General (DG), each of which is responsible for preparing and implementing technical legislation. With regard to design management issues, there are DGs charged with a number of responsibilities of interest including:

- Information, communication and culture.
- Science research and development.
- Telecommunications, Information Industries and innovation.
- Customs union and indirect taxation.
- Internal market and industrial affairs.

With regard to competition policy, the European Commission has prevented large subsidies being granted to companies where it deems this will give national industries an unfair advantage (e.g., the cancelled French subsidy to Renault in 1996). Within the consumer electronics industry, plans by Mme Cresson's government to provide a £200 million subsidy to Thomson were scrapped after opposition by the Commission.

The SEM requires designers and manufacturers to work within a new infrastructure of legal and technical data, with safety standards, testing and certification procedures beginning to harmonise throughout the EU. The onus is on designers and manufacturers to obtain current information, which is essential to achieving competitive design solutions. Achieving such goals requires design teams to have an understanding of the grants available to encourage innovation and how they can be obtained and distributed. But has EU bureaucracy stifled companies attempts to develop innovative design solutions by failing to offer clear advice on procedure?

The Cecchini report

Cecchini (1988) predicts that deregulation of the single market offers potential savings to companies through lower costs from economies of scale. While this opportunity may exist, firms can have difficulty in realising cost savings in some industries due to cultural diversity across the EU and the uneven distribution of wealth amongst Europe's population. Climatic and linguistic differences, coupled with a wide variation in income per head across Europe, presents real difficulties in marketing common products and services throughout the SEM. For example, consider the cultural differences between Northern Sweden, which lies within the Arctic Circle and Greece which lies on a latitude of 40 degrees receiving a Mediterranean climate.

The wide variation of income per head between these two countries also presents difficulties in marketing common products and services throughout the EU. Consumers in different parts of the EU will interpret 'innovatory products' in different ways. Recent work by the Henley Centre, demonstrates the influence which consumer perceptions of quality, performance and identity can have on product sales. Design can be characterised as a people business and for companies aspiring to design pan-European products, it is of crucial importance to identify the type of people who are likely to be targeted as customers (Laurent, 1983). Companies aiming to develop and market pan-European products are faced with disparate labour pool structures, varying consumer perceptions and preferences, all within a diverse spectrum of cultural backgrounds (Price Waterhouse and the Cranfield School of Management, 1991). Research by the Henley Centre (1992) has shown no clear emergence of a 'Euro consumer'. It is likely that European consumers will belong to one of 6 groups, shown in figure 1, characterised by social traditions reflecting centuries of culture.

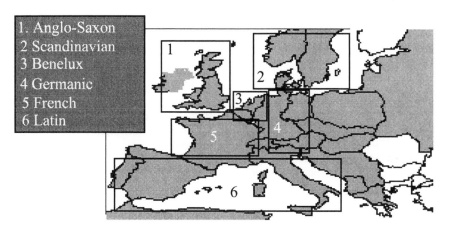

1. Anglo-Saxon
2 Scandinavian
3 Benelux
4 Germanic
5 French
6 Latin

Figure 1 Consumer profiles across the EU.

Euro-consumer convergence

Diversity in management styles, social and cultural traditions and economic differences all present barriers to the establishment of a Euro-brand. Dudley and Martens (1993) argue that in cross border marketing a brand must convey meaning and be socially acceptable, which can cause problems in translation. Since 1992, many companies have been expecting the emergence of a 'Euro-consumer', encouraging growth in Euro-branded goods. However, consumer convergence has been limited. Although forms of cultural expression are becoming increasingly international in character, national loyalties are still evident in the products people purchase. Much of what is common across the EU stems from US culture, helping to underwrite the success of American companies in the SEM.

Research from the Henley Centre points to the difficulties arising from linguistic diversity. In some instances, the lack of fluency amongst EU consumers has given opportunities for companies to use foreign languages to bestow their products with positive connotations associated with foreign countries. Although Europe has a relatively well travelled and well educated society, there has been no post 1992 large scale professional migration and therefore no mass language learning programme. A survey conducted by the Henley Centre and Research International in 1992 also argues that while many consumers are prepared to accept Europeanisation, relatively few feel European, including younger age groups. They conclude; " The positive features which have been associated with Europe in some regions will be lessened by a deterioration in the economic climate, greater competition and a widening membership of the EU. As a consequence, the ability to gain competitive advantage by using European symbols will decline".

The extent to which companies can make use of common symbols or personalities in their promotional strategies will be limited by variations in emphases across the EU and by a continuing national bias in television. All member countries of the EU have relatively highly developed economies with many of the obvious consumer and industrial products being commonplace. Clearly, for some companies, Euro-branding will have little significance, while for others with a well researched product, supported by substantial investment in promotion, long term cost savings and a competitive advantage are possible rewards. In the context of consumer profiling, three established routes could be extended to provide a platform for constructing a consumer based product strategy:

- Market research across a broad span of information available to a company from European institutes and professional bodies.
- Test marketing of existing or modified products in targeted EU countries.
- Acquisition or corporate venturing.

THE SEM - OPPORTUNITIES FOR INNOVATION

The advent of the SEM has presented a concoction of positive and negative influences to European and many foreign (notably US) companies. The potential impact on competitiveness and profitability cannot be ignored. For European firms, some will identify new markets while others will go out of business in the face of increased competition. Protectionist policies and trade barriers operated by national governments will disappear. Industry restructuring may affect an established supplier base and customer pool. For foreign companies, some will exploit new sales opportunities within the SEM while others will encounter increased global competition from stronger European industries. The potential dangers and benefits of operating within a single market are summarised in figure 2.

Realising economies of scale

Pursuing economies of scale has created over capacity in some sectors of industry, encouraging manufacturers to redefine their market or develop a specialist niche (see figure 3). Potential economies should be initially assessed by an analysis of the company's cost structures in manufacturing, technology, transport and distribution logistics. The investment risk to achieve economies of scale should be reviewed in respect of the planned economy over a realistic time-scale. Selling opportunities also need to be predicted to increase sales, increasing capacity otherwise there is no benefit in increasing capacity.

Funding innovation

Cecchini estimated that a fully integrated market could realise an increase of 50% in private sector investment in science and technology being directed towards the competitiveness of European companies. Within the EU, levels of spending on research and development vary substantially, with only Germany coming close to the US level of 2.8% GDP or the Japanese figure of 2.5%. The Latin countries and Ireland all figure in the bottom of the league on R&D spending, with Greece at 0.4% of GDP and Portugal at 0.5%. This is reflected in the different levels of manufacturing sophistication throughout the EU. The SEM gives an opportunity to access sophisticated manufacturing processes and techniques, if these are superior to domestic industries. The European business objectives set by a company, will determine the profile of its research and development strategy. This in turn will influence the composition of manufacturing and marketing criteria in the development of a group of strategies for the execution of new product development programmes. The European Commission has authorised business consultants to report on a range of leading industries over the last decade and these reports can be obtained from one of the EU's cross border information centres (see figure 4).

Research and Development is the largest cost centre within many manufacturing organisations and a company's aspirations may be restricted due to limited availability of resources. To tackle this problem, the EU operates a number of funded schemes to promote research and development which would normally be beyond the resources of individual member countries. Such programmes are implemented through two main routes:

- Joint Research Centres (JRCs) in Germany, Netherlands, Italy and Belgium.
- Co-ordination of national programmes through contracted independent research laboratories across the EU.

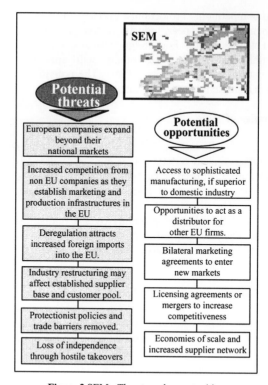

Figure 2 SEM - Threats and opportunities.

Figure 3 Strategic options for product / market development

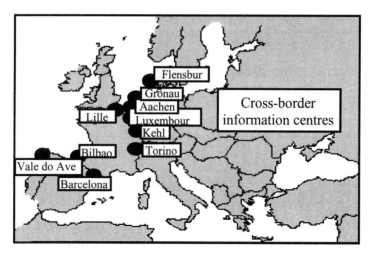

Figure 4 EU information centres.

The management of specific research initiatives is coordinated by several committees. A few examples of principal committees are shown in table 1.

Table 1 EU controlled Research Committees

Committee	Area of specialism
IRDAC	(Industrial Research & Development Advisory Committee)
CREST	(Committee for Scientific & Technical Research)
CODEST	(Committee for the European Development of Science & Technology)
ESPRIT	(European Strategic Programme for Research & Development in IT)
RACE	(Res. & Dev. in Advanced Communications Technology for Europe)
BRITE	(Basic Research in Industrial Technology for Europe)
EURAM	(European Research in Advanced Materials)

There are other programmes, such as EUREKA (covering industry related R&D) which have operated throughout Europe outside the control of the EU. This highlights one of the main difficulties for companies seeking advice on developing innovative design solutions; which agency or committee is most appropriate to their needs? Levels of spending on R&D throughout the EU have previously been linked to the economic performance of individual countries, but this relationship is diminishing in relevance as European economies converge. Based upon figures released in February 1998 (see table 2), only Greece failed to meet the targets for Economic and Monetary Union (EMU) set at Maastricht in 1992. Despite some observers claiming that Germany, France and Italy have had to resort to 'alternative accounting' measures, eleven member states look certain to form a 'broad' based EURO by January 1999.

Table 2 EMU - Economic performance indicators.

	Deficit/ GDP	Debt/GDP	Inflation
Target	3%	60%	2.7%
Austria	2.5	66.1	1.2
Belgium	2.1	122.2	1.5
Denmark	0.7	64.1	2.0
Finland	0.9	55.8	1.2
France	3.0	58.0	1.3
Germany	2.7	61.3	1.5
Ireland	0.9	67e	1.2e
Italy	2.7	121.6	1.9
Luxembourg	1.7	6.7	1.4
Netherlands	1.4	73e	1.9
Portugal	2.5	62.0	1.9
Spain	2.6	68.3	1.9
Sweden	0.4	76.6	1.9
UK	0.4	53.4	1.9

e = Estimated   Source: Reuters/Eurostat

However, Europe has fallen behind the US and Japan in the sector of high technology. Measured as a percentage of the GDP, Japan spends 2.5 % on R&D of which over 1.5 percent is funded by the Japanese government. This compares with an average of 1.9% spending on R&D in Europe and the US figure of around 2.8%. An ex Vice-President of the European Commission, Karl Heinz Narjes (1995), has highlighted this gap;". . Europe still has a significant gap to fill in a number of high technology sectors. These are precisely the sectors which will determine our competitiveness in the decades to come - Information technology, biosciences and new materials. If Europe does not succeed in reducing its technological dependence in these fields over the next few years, the Communities internal market will also remain fragmented."

CONCLUSION

In summary, when setting realistic objectives for the design and manufacture of innovatory products, a wide range of factors relating to social, economic and technical diversity within the SEM will influence the policy. Deregulation has presented opportunities and posed threats to European businesses seeking to sustain a competitive edge in an international marketplace. A company seeking to use technological innovation in new product development must take on board a range of factors which will influence a company's decision concerning the appropriate technology and the wider implications of its use (see figure 5).

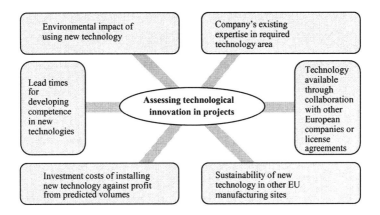

Figure 5 Factors influencing adoption of technological innovation.

In post war Japan, the Japanese government established a Ministry of International Trade and Industry, (MITI), which built up long-term relationships with Japanese companies mainly in the pro-competitive arena, by adopting a number of strategies to create a favourable economic and industrial environment for growth. The MITI was orientated towards exploiting the most propitious international opportunities available by encouraging research and development particularly in high technology companies. The EU has no equivalent of the MITI. European companies have to deal with a variety of sources for advice and funding, creating 'overlaps' and 'gaps' in the system. There is no single telephone number for help. This paper has examined the impact of these factors upon the ability of EU member states to manage new product innovation and has attempted to identify the link between innovation, demographics and cultural diversity. It would seem timely to press for a more coordinated and less bureaucratic approach to the control and dissemination of funding towards the promotion of technological innovation in product design.

REFERENCES

Bacon, F. and Butler, T., 1981, *Planned Innovation*, (Ann Arbor: University of Michigan).

BS 7000, 1989, *A Guide to Managing Product Design*, (British Standards Institute).

Cecchini, P., 1988, *1992; The Benefits of a Single Market,* (Wildwood House).

Davies, H., 1993, The impact of Competitive Structure and Technological Environment on Design Management: a case study of the UK touring caravan industry, In *Design Studies*, Vol. 14, No 4, pp. 365-378.

Dudley, J. W. and Martens, H., 1993*, 1993 and Beyond: New Strategies for the Enlarged Single Market*, (London: Kogan Page).

Henley Centre for Forecasting Ltd., 1992, *Frontiers*, (London).

Hollins, W. and Hollins, G., 1995, The Management of Concurrent Engineering and Total Design. In *Co-Design Journal*, No. 03-95, pp. 22-27.

Hughes, C., 1995, More power through new approaches to product development. In *Co-Design Journal*, No. 03-95, pp. 58-63.

Laurent, A., 1983, The Cultural Diversity of Western Conceptions of Management. In *International Studies of Management and Organizations*.

Narjes, H. J., 1995, Ex Vice-President of the European Commission. Speech to the European Commission, Brussels.

Nelson, R. and Winter, S., 1982, *An Evolutionary Theory of Economic Change*, (Cambridge, Mass: Harvard University Press).

Oakley, M., 1990, *Design Management, A Handbook of Issues and Methods*, (Oxford: Basil Blackwell).

OECD, 1981,   *The measurement of scientific and Technical Activities* (Frascati Manual), Paris, Organization for Economic Cooperation and Development.

Papanek, V., 1971, *Design for the Real World*, (London, Thames and Hudson).

Potter, S. and Roy, R., 1986:, *Design and Innovation, Block 3: Research and Development; British Rail's Fast Trains*, (Milton Keynes: Open University Press).

Price Waterhouse and the Cranfield School of Management, 1991, *Report on International Strategic Human Resource.*

Type and Identity: a scientific approach to graphic design research

Ian Montgomery, Kenneth Agnew, and Brian McClelland

INTRODUCTION

We look at objects, images, buildings, typography both cognitively and affectively. We assess, categorise, associate, dissect, reject, accept - we make judgements. In our society which expects quality, and usually gets it, design directors are now looking toward styling and corporate identity to keep them competitive in an increasingly hostile marketplace. In manufacturing, product design and typography are fundamental elements of many corporate identity programmes - almost all useable manufactured products contain some element of typography. Each product and typeface has character, they can usually be described using a range of words which we associate with feelings. 'Sets' of words (descriptors) are used to describe the many facets of a product, service, environment or behaviour - all of which constitute an identity. This paper investigates corporate identity, product and typographic character and outlines an experimental method for assessing perceptions of 'appropriateness'.

IDENTITY

It is becoming more difficult to differentiate between the mainstream generic amorphous automotive shapes, the portable CD players, the 'slate grey' televisions or the corporate seafront furniture of the late 1990s. The move toward corporateness and the family of products, while strengthening the overall image of the company has produced individual products which are so corporate that they lack a strong individual identity. The adoption of the corporate typeface as an identifier for an entire family of products has inevitably quoshed the idea that a great deal of the decisions taken by the designer should be based on the concept of appropriateness.

Olins states: "...products from the major competing companies around the world will become increasingly similar. Inevitably, the whole of the company's personality, its identity, will become the most significant factor in making a choice between one company and its products and another" (Olins, 1989). The desire to project an image of quality, strength, efficiency, tradition, excellence has compelled mass manufacturers to engage in broad sweeping visual statements. The characterless number coded product is no longer confined to the living room,

Rover, Mercedes, BMW, Peugeot and Volvo have all taken this route. The concept of communicating product through numbers like 620 with the important 'i', 500 with 'SEL', 540 with 'M' has deprived some fine mainstream products of character. As Kim Basinger said in the recent television advertisement for Peugeot "What's a 406"?

As a tool for transcending international boundaries the use of numbers is a corporate success - the importance of choosing an appropriate style of typeface to reflect the ethos of these companies has its allegiance in the 'corporate' rather that the 'individual' product camp. Numbers, as applied to products, tend to be unambiguous. In the case of Rover and BMW the same number style is used for all cars - arranged in a range of sequences to denote various types, sizes and pricing structures. The identification numberstyle used for a basic Rover 200 is the same as that of the very much larger and more luxurious 825 - both cars are very different in size, style and price - but are 'badged' as corporate products.

The contemporary concept of corporate identity fails to take into account the effective communication of form types and product character through appropriate typography. As the 'family' of products continues to grow and niche markets are developed so the need to develop more 'individual' namestyles is vital. The communication of a product type can be discerned often by a small arrangement of letters or numbers. The style and delivery of these identifying symbols tell as much of the product 'story' as intricate body styling details yet they seem not to be treated according to the character of the form - but represent some far off concept of corporateness.

CHARACTER

In all stages of the design process descriptors are used to describe feelings of shape and form - they assist in the communication process within and between groups of designers, designers and clients and are used to assist product marketing. Just as products are perceived to have a certain character, a feeling, a richness of style so the perception of all areas of design are governed by the same principles but within the parameters of their varying disciplines.

The description of character is used by designers at all stages of the design process and beyond - from concept to realisation, in manufacture, use and obsolescence. Visual cohesion through all elements within a design should be a paramount concern for designers if the character of individual forms is to be effectively communicated.

Intended character

The verbalisation of form and visual communication as used prior to and during the visualisation stage contributes to the development of character in forms. Michael Tovey referred to the panels which constitute an automobile form as: "...surfaces which meet at intersections, are filleted, or blend smoothly into each other. The forms may be hard, or rigid, or soft and flowing or a judicious combination of both." He goes on to state: "Surfaces which are heavily curved in both planes would be described as 'loaded' or 'meaty', very flat surfaces or those with crisp edges or features might be described as 'taut', as having some tension" (Tovey, 1992).

Complex forms, by their nature, require sets of words to make explicit those characteristics which they possess. It is important to differentiate between the use of groups of descriptors which are clumped together to describe a form and the holistic approach used in developing or communicating a concept. Design intentions are sometimes referred to in vague, yet valuable, descriptions which may encompass boundaries across and between cognition and emotion. Designer Bjorn Envall described his intention for the Saab 900 series: '...to produce cars with fine ergonomics and with looks which, if not beautiful, have a subtle long-term fascination' (Envall, 1985).

According to Darrell Behmer (1998), chief designer of the new Ford Cougar: "Cougar is bold, confident and optimistic. We wanted it to satisfy the emotional instinct and exude a certain strength – underlining the car's true driving character while maintaining an air of sophistication. Functionality was a guiding principal. Our goal throughout development was to create a car that was sleek, with classical coupé proportions" (Behamer, 1998).

The use of descriptors is incorporated into this statement of design intent, creating a myth, an aura around which the philosophy of the company and the ethos of the car can be communicated. The use of emotive descriptors such as 'bold', 'confident', 'optimistic', 'exude strength' infer that forms possess human qualities and must be considered in the context of both corporate and individual form identity. It is a policy statement - a corporate strategy and a statement about the characteristics of a form rolled into one.

Keith Clements, Chief Designer of the Ford Ka, stated: "What we were pushing for was a design that was simple yet distinctive, sophisticated yet futuristic." (Clements, 1998). The style and design of the logotype shares some common characteristics with the manufactured form.

Perceived character

Krampen (1989:1) uses descriptors in experiments to: "understand what the meanings of buildings and designed objects might be." He defines 'denotative meaning' as: "...what different types of buildings or objects afford in terms of different functions." Krampen goes on to state: "Connotative meanings result from the fact that every architectural or design problem can be solved in (many) different ways." In an experiment he used two different types of door handles to show the relationship between the product and its perceived connotative meaning:

handle type	connotative meaning
light alloy	simple, sober, clean, linear, slick, unobtrusive, modern, pleasant, economical, cheap, modest, cold
non-ferrous (*with decorations*)	old-fashioned, striking,, ornamented, decorative, luxurious, playful, flourished, craftsmanlike, warm, soft

In the specific instance of the Aston Martin DB7 Bayley (1994) compares its character to that of the content of a 1920s British thriller. He says: "If you were to lock a control group in a white-tiled room and make them do free-association tests under painfully bright lights, they would come up with terms that sounded like a semiotic analysis of a John Buchan thriller: handsome, rugged, tough, honest, elegant, fast, masculine. Its shape is ravishing and authentically beautiful: it has perfectly judged proportions, nice details and sumptuous curves" (Bayley, 1994).

The use of metaphors to describe engineering forms is commonplace. Richard Seymour (1996) makes the observation: "Ka is an egg, with a Grace Jones haircut" he goes on to say: "Why do I get the feeling when I look at it that Ka is a bit of a cross-dresser, a plain woman with too much rouge" (Seymour, 1996). Just as Seymour has used direct and vivid descriptors to communicate the character and essence of a visual form so other areas of design are judged and discussed.

Coates (1978) has modelled a semantic differential (Osgood, Suci and Tannenbaum 'The Measurement of Meaning' 1957) along with relevant antonyms to describe the range of perceived descriptors which encompass the German versus the American 'feel':

judgements of German cars   **judgements of American cars**

austere	— — — — —	lush
simple	— — — — —	complex
rational	— — — — —	emotional
smooth	— — — — —	rough
controlled	— — — — —	accidental
ordered	— — — — —	chaotic
plain	— — — — —	ornate

Krampen (1989:2) has carried out experiments in which he investigated effective meanings of four different car body designs. Overlap between the various designs show the use of common descriptors:

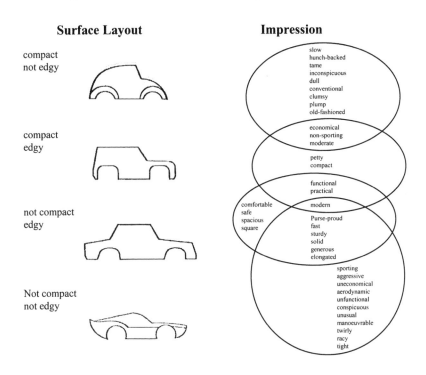

TYPOGRAPHY AND PRODUCT DESIGN

According to Walker, Smith and Livingston (1986), "When graphic designers select a typeface that is appropriate for the concept that their design represents, they make use of a valuable opportunity to enhance the overall impact of the design". In the instance of product graphics where appropriateness straddles both corporate philosophies and form considerations many poorly communicated and ill-considered choices are in evidence. An example of appropriate choice of typeface for Tranquillity Eau de Toilette and inappropriately used for power tools.

Tranquility
Eau de Toilette

Strike Force
Power Tools

Just as forms are deemed to possess an inherent character so Walker, Smith and Livingston believe: "The notion that a typeface can have a character that makes it appropriate for a particular concept, such as a business organisation, implies that it shares with the concept those features that collectively define its own character. Consequently, any attempt to understand the principles underlying typeface appropriateness must begin by identifying the features that a typeface may share with such diverse things as organisations, perfumes, food products and music" (*ibid., p.30*). This concept of 'appropriateness' is further referred to by Walker et al as: "The question of what makes a typeface, or any other visual element of a design, appropriate is of central concern to designers" (*ibid., p.29*).

Crozier (1994) is more vague about the role of typography within product perception. He believes in the corporate ethos and is more concerned with corporate communication than form values. He states: "The success that a typeface has in, say, providing corporate identity may be due not so much to its formal characteristics but to some shared, if unconscious, meaning, shared between the design and the image of the company or product that is identified" (Crozier, 1994).

A range of scientific experiments have been designed to investigate levels of appropriateness and to discover if, and to what extent, perceived relationships exist between typefaces and forms.

THE SCIENTIFIC APPROACH

To set the scene, the scientific approach to aesthetic judgements does not hold much ground with many commentators. In an article on car personality Bayley states: "There is no basis in aesthetics for a scientifically valid law" (Bayley, 1995).

However, to provide evidence of associations existing between two different specialisms within design it is necessary to conduct research based on fact - not, as so often happens in design, on conjecture.

Since product/typographic interaction and interface is an area of research which has been incompletely explored much of the experiment design work has been self-generated. Consequently the choice of experimental constraints and objectives were made with particular care in order to create limited numbers of variables and which refer, wherever possible, to the existing scarce references. Accurate record-keeping, effective and appropriate documentation of experiments and graphic presentation of data were essential if clarity was to be achieved in a positive, negative or null hypothesis situation.

EXPERIMENT 1

This experiment was designed to establish, through a simple method, a list of fonts associated with forms and conversely forms associated with fonts. A range of pilot studies using a similar method to that outlined below had been completed using a small number of participants. A simple two-part matching method was utilised to gather information which was analysed, reported and presented as part of the wider research project. The experiment was conducted with two different groups of 30 participants.

Method part A

A matching technique using 14 font variables and 14 individual forms (shown to participants one at a time) was used to gain first, second, third and fourth preference allocations by a group of thirty participants.

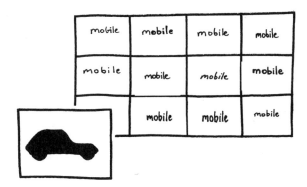

Figure 1a

Method part B

A matching technique using 14 form variables and 14 individual fonts (shown to participants one at a time) was used to gain first, second, third and fourth preference allocations by a group of thirty participants.

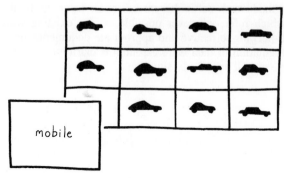

Figure 1b

Experiment format

The design was selected as an effective and straightforward method of providing matching lists. The experiment was a paper-based matching test under standard experiment conditions. In both parts of the experiment fourteen dependent variables were positioned onto a wall. Independent variables were provided as hard copies. There were 30 participants.

Purpose

The two sets of experiments were conducted using a random sampling of participants who were not high-level visual analysts to provide information of basic visual associations between form and typography.

Data received was reviewed and evaluated for the purpose of:
- establishing lists matching form and typography and vice versa
- establishing theories about links between typography and form.

This set of experiments was designed to be of use in a number of ways:
- evaluate and develop the method effectiveness
- to test a range of statistical methods in evaluating the results
- to further determine the extent to which associations exist between form typography.

The experiments yielded some interesting results. There was evidence of consensus being reached in assigning forms to typefaces in a number of instances. Consensus between participants was also evident in the non-assigning of forms to typefaces in other instances. Some results are shown here. Results show that in a number of instances statistical peaks and troughs match between one group of participants and the other.

In Part A there was consensus between groups 1 and 2 regarding the five highest 'scoring' typefaces. These were:

Font 1

Font 6

Font 9

Font 11

Font 12

In Part B there was consensus between groups 1 and 2 regarding the four highest 'scoring' (those most allocated in a first preference vote by participants) forms. These were:

In Part B there was consensus between groups 1 and 2 regarding the four highest 'scoring' (those most allocated in a first preference vote by participants) forms. These were:

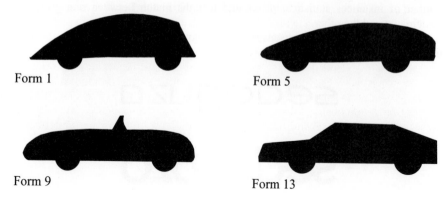

Form 1   Form 5

Form 9   Form 13

Of particular significance is the matching of Form 5 with Font 11 and Font 9 with Form 6 and Font 4 with Form 9.

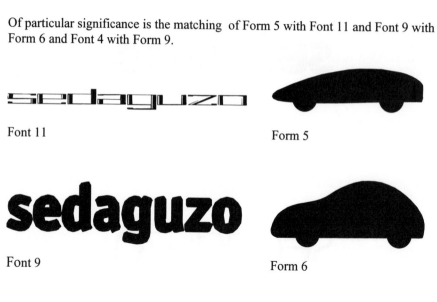

Font 11   Form 5

Font 9   Form 6

Font 4   Form 9

The examples outlined above show elements of visual similarity. Where single silhouettes of forms were matched with one specific typeface similarities in shape and style can be seen. In the comparison of Font 9 with Form 6 the same descriptors could have been used to capture the character of both. They could both share descriptions like: 'bold', 'rounded', 'soft' and 'bulbous'. Further experiments have been conducted using 'semantic differential charts' and other techniques in the development of some theories set out in this paper.

The use of a range of descriptors to encompass a feeling of character and identity must be extended to aspects of individual forms if the identity of products is to be preserved. The need to share visual cues between graphic and formal elements of design must be addressed if products are to be developed in a progressive, aesthetically individual and wholly appropriate way. The need to reassess the approach toward, and application of, visual information to products must be a primary consideration for engineering designers, marketing executives and design educators.

REFERENCES

Bayley, S., 1994, What makes a British car British? In *Car*, May, p.78.

Bayley, S., 1995 The most important thing a car can have is personality. In *Car*, April, p.71.

Behmer, D., 1998, *Ford Cougar*, Ford Press Pack promotional brochure, p.5.

Clements, K., 1998, *KA Edition 1*, Ford promotional brochure, p.1.

Coates, D., 1978, The European look spreads - or is it just good design? In *Industrial Design*, Sept/Oct, p.36.

Crozier, R., 1994, Manufactured Pleasures, (Manchester: Manchester University Press), p.158.

Envall, B., 1985, in Bayley, S., *The Conran Directory of Design,* (London: Conran Octopus Ltd), p.122.

Krampen, M., 1989, *ibid.,* pp. 136 – 137.

Krampen, M., 1989, Semiotics in Architecture and Industrial/Product Design. In *Design Issues*, Vol. V, No.2, p.134.

Olins, W., 1989, *Corporate Identity,* (London: Thames and Hudson), p.9.

Seymour, R., 1996, Over The Edge. In *Blueprint,* October, pp.23-24.

Tovey, M., 1992, Intuitive and objective processes in automotive design. In *Design Studies,* Vol.13 No.1, p.23.

Walker, P., Smith, S. and Livingston, A., 1986, Predicting the appropriateness of a typeface on the basis of its multi-modal features. In *Information Design Journal,* Vol.5 p.29.

Empowering the Design Team: a multimedia design resource to facilitate the capture, retention and reuse of knowledge acquired during product development

S. E. Phillips and J. T. McDonnell

We report on the use of an issue based representation to structure large amounts of product design information and construct a multimedia based design resource. In a detailed field study of the product design process we have addressed the problems of eliciting tacit design knowledge, capturing design rationale and representing the resulting information in a form which is accessible to designers and useful for reference in their future work.

We give an account of how the designers in the field study reacted to making use of the tool as an information resource, and categorise the issues that were raised. We show how we are extending the notation for recording the design rationale to cope with the pragmatics of real design processes. We describe the way the resource needs to be structured to give the designers useful access to their previous design decisions and product development influences to support personal and team reflection, company knowledge management and the maintenance of brand direction.

INTRODUCTION

It is often claimed that tools to assist design based organisations to capture and retain the knowledge and reasoning behind their product development will greatly advantage these companies. However it is difficult to establish what the tangible benefits might be without empirical studies in which knowledge based resources are made available to a design team so that the value of the resource can be tested in real use situations. Designers often get inspiration by looking through old designs and samples and many diverse sources are used to obtain design information. Much of the material referred to and created during the design process forms an inherent part of the rationale for the products eventually developed. This information may take many forms. As well as including data about previous products, competitors' products, sales information and marketing directions, there may be video sequences and story-boards containing pictures and media samples intended to illustrate the design proposals.

We have suggested that a multimedia platform in which information can be organised in a structure which reflects the design decision making process would be a valuable resource in design driven industries (Phillips, 1997). By retaining information on design decisions and influences, the departure of members of the product development team would be less damaging and new personnel would be provided with an insight into the rationale behind previous design decisions. This is particularly important when designing branded goods where image consistency is all-important. In design orientated companies it is often common for many prototypes to be created during the product development cycle, usually only one of which will be selected for production. The prototypes that are not selected are discarded and the research required for their development is also lost. If this data were retained, it might prove valuable when designing other products. The resource we propose would also be valuable to production and marketing activities (Figure 1).

Most companies find it difficult to preserve information from the product development cycle, it is neither captured nor retained for future reference. Even if there is storage space, lack of structure, loss of context, and changes in personnel soon reduce the usefulness of information about previous design work. Furthermore, product designers are not traditionally required to be explicit about the rationale behind their designs and therefore a substantial amount of the information that they use during the design process is not shared with others or formally recorded. To explore the practicality of building and using a multimedia design knowledge base we set out to capture the design rationale for a Spring/Summer 1998 product range for an American sportswear company. A prototype knowledge structure was constructed, using the data collected from the field research, and based on an existing design rationale notation. This gave the designers a resource with which to experiment. Through our discussions with them about this experience we have been able to establish how the notation needs to be extended to cope with the practicalities of real design processes and what sort of structure the resource needs to have to be useful to them in future design activity.

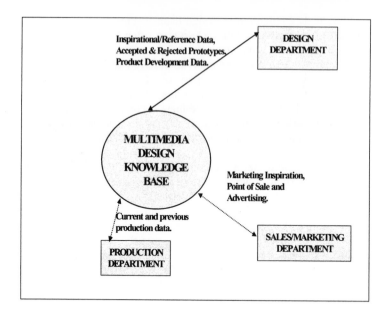

Figure 1 Organisational interactions with a multimedia design knowledge base

CONSTRUCTING THE KNOWLEDGE BASE

The idea was that the knowledge base should store all the information relating to the products being designed and encapsulate the reasoning behind their development. It should effectively capture the design rationale over a period of time allowing the design team to track progress season by season and show the evolution of the products.

Several well-documented methods for representing design rationale were evaluated and we chose a notation developed from the original proposals made by Kunz and Rittel (1970) now further developed and commonly referred to as an IBIS (Issue Based Information System) notation (Moran and Carroll, 1996). The main advantage of this notation is that it has been developed to be used alongside decision making, as the deliberation actually takes place. Since we needed the capability of associating multimedia data sources with the maps of the design decisions we were creating we used a tool which enabled us to link files of any format to nodes in the maps. The final knowledge structure contained approximately 200 image files, 4 video files, 2 audio files, and 30,000 words of design meeting transcripts. A small fragment of one of the IBIS maps can be seen in Figure 2.

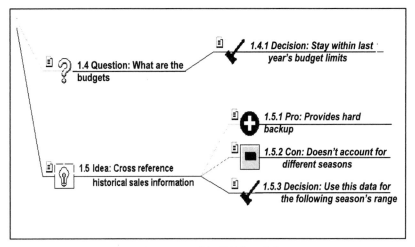

Figure 2 Fragment of the IBIS knowledge map

Each node contains explanatory text and may be connected to other reference material such as other IBIS maps, graphic, video or audio files. The different node types and the totals of the nodes and links contained within the 13 IBIS maps of the completed knowledge structure can be seen in Table 1. This knowledge structure was installed at the designers' premises for assessment of its effectiveness and limitations.

Table 1 Totals of the different node and link types within the decision maps

Node Types	Totals
Questions	14
Ideas	134
Arguments	30
Decisions	71
References	201
Notes	17
Map views	13
Total Nodes	**480**
Total Links	**471**

To get valid responses from the designers it was thought essential that the resource they were given to work with was realistically complex, and as complete and accurate as possible. To achieve the necessary complexity and detail in the knowledge base, all the design meetings were attended for one whole product development season. All of the formal design meetings were fully transcribed, and further visits were made to the company to interview members of the product development team to collect other background material for the knowledge base.

ANALYSIS OF THE KNOWLEDGE STRUCTURE

A series of meetings were held with those members of the product design team who had made use of the knowledge structure, we summarise some of the main finding below.

Capturing design rationale in its entirety

Concern was voiced regarding the impossibility of capturing all the design decisions, despite the fact that as far as possible all the formal design meetings had been attended. It might be expected that all important decisions would be shared or discussed in the formal design meetings, however, this is clearly not always the case. Issues may be discussed between some of the design team on the way to meetings in the car or in other non-formal settings. Furthermore, urgency sometimes dictates that decisions have to be made outside formal meetings. There may be no time to consult or inform the rest of the design team about decisions made by one member of the team with regard to issues that fall under that particular person's area of responsibility. These are practical constraints that we must accommodate. We propose that these shortcomings can be overcome to some extent by holding a session for reflection at the end of the design process during which the decision structure is reviewed, 'corrected', annotated and rationalised. We have estimated that this task would take no more than a couple of hours and would ensure that the design rationale structure accurately reflected the important outcomes of the design process. In itself this session would be useful as a review for the design team of the design process and the goals that had or had not been achieved.

It was also clear that the maps would contain more detailed and accurate knowledge had they been created by the product development team themselves, as opposed to a knowledge engineer. Then, however the product design team would need to find the time and the motivation to do this, it certainly was not feasible with the tools we were using to structure the data from the design meetings. This is an area that requires further investigation.

There are also questions to be addressed over the amount of tacit knowledge that is brought on bear on understanding the rationale as it has been recorded. With visual data such as story boards and inspiration videos it will not be obvious to a lay person what is being conveyed, this is part of the skill the designer and the unexpressed culture and context within which the team operates effectively. Where there is a debate on a specific visual and aesthetic issue, it must be considered that the data will be looked at in an historical context in the future. For example, if there is a discussion involving the use of particular fabric types, investment in clarification such as the inclusion of visual illustrations will demonstrate what the fabrics look like, and what their different types and applications are at that time. Cultural perspectives of the time must be added to the data in order for it to retain its meaning in future years. Identifying exactly what will be needed is ultimately impossible. However practical, experienced designers can probably make intelligent guesses that will help their successors.

Extending the IBIS notation

Certain issues were raised due to limitations with the IBIS notation we used. This was restricted to mapping the rationale in terms of nodes associating questions, ideas, arguments for and against, and decisions. Our approach to tackling the weaknesses identified has two main aspects. The first is to devise additional node and link types to make the representation richer. The second is to introduce a distinct reflective stage that follows on from the design process during which the maps are in effect 'tidied up' to make them valuable for future reference.

The representation we used initially required decisions to be concluded very cleanly fully resolving all the issues and ideas along the development path. In practice this does not happen, therefore as they stand the maps mislead the user into thinking all the issues in a line of development have been addressed, when in actual fact they have not. Sometimes there is a very accurate representation of the line of reasoning, which leads directly to a full resolution of all the issues and ideas in the line. At other times, a decision may be made that only partially addresses the line of reasoning, the representation must be able to show this. Where decisions are made on the maps it is not clear whether all the ideas or just some of the ideas along a line of reasoning have been solved by a decision. We propose to distinguish between these two cases by using the term 'decision' to refer to decisions which accommodate some of the ideas along the line of reasoning but which, for practical reasons disregard others and the term 'resolution' to refer to decisions which accommodate all the ideas along the line of reasoning as shown in Figure 3. This allows us to retain the simplicity of presenting the reasoning in linear form.

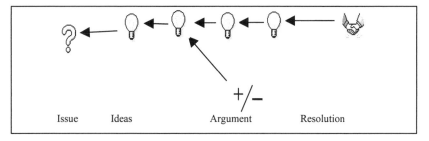

Figure 3 Adding the new node type: resolution

The design team was keen differentiate decisions according to which department or perspective they were coming from (e.g., Marketing, Sales, Management, Production, Design). For this we propose to use colour coding on decision and resolution nodes. It was also clear that some temporal aspect to the representation would be valuable so that when reflecting on the design process it would be possible to see in what sort of time frame decisions were made. Furthermore, the team would like to use the knowledge maps to assess whether they are reaching their targets within the expected time frame. A simple but adequate solution for these purposes is to time and date stamp all the nodes as show in Figure 4.

The design team felt that until construction of knowledge maps becomes an integral part of the design culture, there would be limitations in the information that a user could gain through a quick glimpse at a map. It would be important to ensure that once the final products are determined they are clearly and directly linked back to the initial ideas to which they relate.

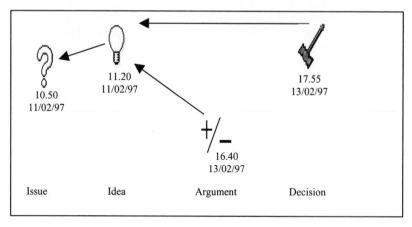

Figure 4 Time and date stamped nodes

The simplest way of achieving this is to link retrospectively the key ideas nodes to graphics files of the end product they affected. Such a link from an ideas node can be shown by colour-coding which alerts the user to a retrospectively added association, as shown in Figure 5.

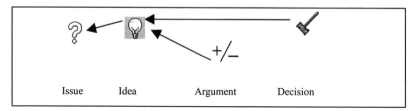

Figure 5 Colour shaded idea node denoting a link to the finished product file

Another interesting observation, which comes from carefully recording what actually goes on during design meetings, is that many of the ideas that are raised appear to remain unresolved. During retrospective editing decision nodes that are omitted from the maps need to be added and connected to the ideas to which they relate. Otherwise it would appear when looking back in the future that some ideas have not been resolved and decisions have not been made, when actually they have been addressed somehow even if by being swept aside temporarily in the need to get on with other more pressing concerns. We propose that all these 'tidying' up operations that take place retrospectively should be flagged as such, for example by adding an "R" flag to any node added at the retrospective stage, as in Figure 6.

Sometimes decisions are not implemented even though they are discussed during the design deliberations. A new node type called 'not actioned', linked to decision nodes, will be used to annotate the map at the end of the design cycle to ensure the information on the map is up to date. Text should be added to the node explaining why it was not actioned. Where ideas do not have decisions linked to them these should be added retrospectively, flagged appropriately, otherwise if no decisions or resolutions are associated with an idea it can be assumed it was not heeded.

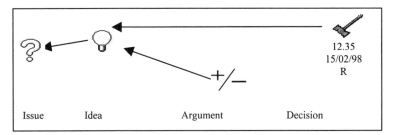

Issue Idea   Argument   Decision

Figure 6 Adding an 'R' flag to retrospective nodes

Supporting reflection/decision making

Several issues were raised concerning the capability of the tool to support reflection on the design process, which in turn would impact on the decision making in the next season's product development meetings. It quickly became apparent that there was a role for the knowledge structure not only as a record of the design rationale but also as a basis for a template to provide a starting point for the following season's design cycle.

The knowledge structure proved useful as a reflective tool for the design team to review the development process for the previous range and to highlight where goals had been achieved. The maps sparked debate within the design team as to why certain issues had remained unresolved or had fallen by the way side. The team had access to the maps as they were being constructed, and while the design process was still ongoing. They noted that the maps were useful to see what goals had been achieved and identify what had been accomplished and what had not, even before the design process had been completed. To a certain extent, at the end of range development, it is possible to review what has been achieved from the initial objectives just by looking at the samples or by flicking through the range catalogue. However the knowledge structure provided an insight into the reasoning behind the decisions being made and allowed identification of the issues that were lost sight of during the design process. Many claims are made about the importance of recording the design process, but in the field of product design little research has been conducted previously to substantiate this.

The knowledge structure provided a starting point for the next season's range planning. It enabled a review to be quickly conducted of what the goals had been at the beginning of the previous season and a new set of goals to be drawn up for the coming season. It became apparent that the decision structure does not change entirely from season to season. Some of the structure will be unique each season, but there is a clear relationship between the structures for each season. The design team wanted to use the structure as a template for the next season's map, to save work in collating the next season's design rationale and enable the new data to be input more efficiently. This suggests that an IBIS-style structure would be used initially to identify a model path for the product development, from which a template could be created, in order to form a basis for subsequent seasons' maps. From each new map a template for the next season's map would be created. This would make the knowledge base, suitably tidied up after a reflective session, a useful tool for product design managers and it would speed up and enhance the following season's range planning. The design team felt it would particularly support strategic planning and decision making at the start of the following season's design cycle.

One reason that not all the rationale was captured in the IBIS maps we constructed was due to the fact that many of the design scenarios understood by the design team, stemmed from previous seasons' development. However if this tool were to be used over a period of time this problem could be overcome to some extent. Maps for progressive seasons could be linked and sub maps of specific issues could also be linked to help address this. If the decision structure were then to be explored in depth for a number of seasons, an understanding of the previous design scenarios and decisions could be gleaned from the structure, making the maps' contents' progressively more understandable to a user, this is demonstrated schematically in Figure 6.

Many other useful ideas were suggested by the design team concerning the linking of other information sources to the knowledge structures for strategic decision support (e.g., linking sales data to the range data, identification of top selling product lines, etc.).

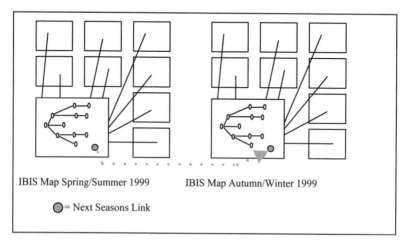

IBIS Map Spring/Summer 1999   IBIS Map Autumn/Winter 1999

⬤ = Next Seasons Link

Figure 6 Linking from one season's knowledge structure to the next

Interface issues

Many issues arose concerning the way information from the knowledge structure might usefully be presented to the designers to support their design processes. The suggestions are the most domain specific of our findings. Most of the requirements can be met by developing a collection of user interfaces to support different ways of viewing information from the knowledge structure. Here we give only a small sample of the ideas our designers suggested.

The maps are currently presented to the users with the final products as the end result of decision paths. For use as a historical record it makes more sense for the user to be presented with the final range of products with the ability to go through the decision making process in reverse to see how they evolved from the original inspiration. Ideally the graphics from each range would be viewed together, or in a sequence of development reflecting the distinct stages of product development, e.g., photos, coloured sketches, and line drawings. Figure 7 illustrates this idea.

When looking at the maps of particular ranges it would be useful to see a gallery view of the range simultaneously enabling users to relate the design deliberations to the products which are discussed as shown in Figure 8.

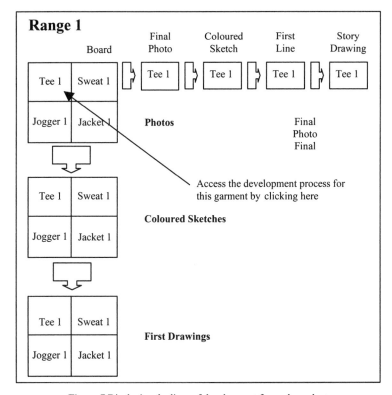

Figure 7 Displaying the lines of development for each product

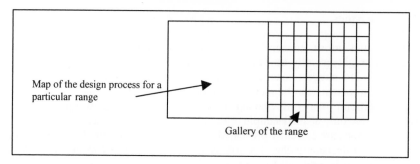

Figure 8 Viewing a gallery of a range alongside a map screen

When viewing an image of a particular product it would be beneficial to be able to view all the other graphics of that product along its line of development, thumbnail images will show the initial sketches, the working drawings and then final colour drawings, to enable the evolution of the product to be seen, as shown in Figure 9. This would facilitate users being able to visually trace back from the final outcome of the design process.

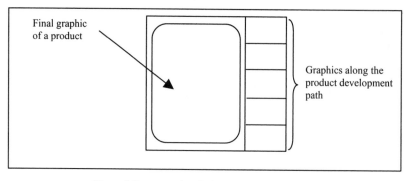

Figure 9 Demonstrating the line of product evolution

Direct benefits

The initial reaction of the design team to the knowledge structure was how well it graphically demonstrated the work conducted and the processes and decisions undertaken during the product development cycle.

A further direct benefit was demonstrated when a member of the design team left the company during the research taking all his knowledge with him. The new team member was to appraise himself of the current design issues using the knowledge structure to help ensure consistency within the next season's range. The dangers to a company of losing its direction in terms of brand image are critical and the knowledge resource demonstrated considerable value in this respect. It was also commented that the resource would be a good tool for reflecting on personal development, by allowing individuals to raise their awareness of their own decision making process and thereby further develop their expertise. The tool also enabled the design team to reflect on how well the external design consultants had fulfilled their roles and how critical or otherwise they were to the process.

It became clear to the product development team, having experimented with the resource, that once the design rationale had been captured over a few seasons the benefits would become more significant. For example, after a number of seasons' data had been collected, it was envisaged that savings might be reaped on paying for external specialist information, such as on 'retro' design data from vintage clothing experts, that once captured would not need to brought in from outside.

CONCLUSION

We have constructed a multimedia design resource structured in the form of a design rationale, based on the IBIS notation, and using existing decision map making tools. We have analysed how designers make use of this. To make the research worthwhile we have ensured that the designers were given a knowledge base constructed from their own design deliberations which is realistically complex, and which was as complete and accurate as possible. This has enabled us to identify how the IBIS notation needs to be extended to be able to express practical design decision making in a competitive commercial setting. It has allowed us to see how many of the inevitable inadequacies of trying to capture design rationale during the design process can be overcome by taking a relatively small amount of time to reflect on the design process and tidy up the record retrospectively. It has led to many useful insights into how information from such a knowledge structure can be presented to designers to support their work.

Currently we are reconstructing the knowledge structure using the extended notation presented here. We are working with maps that now capture the knowledge provided by the designers during a reflection session. We intend to review this revised resource with the designers. We are also building some of the interfaces the designers have suggested. Figure 10 shows the point we have reached and indicates the system architecture we are using.

From our analysis there are some clear limitations which must be overcome if an effective tool for capturing design rationale is to be developed and we are continuing, where feasible, to address each of these. However our work also demonstrated many 'direct' benefits of capturing design rationale using the IBIS notation in the domain of product development.

More research is required in order to assess the feasibility of a design-oriented company allocating adequate resources into retaining product development rationale over a substantial period of time. Nevertheless, it is apparent from the research that if a company invests the time and money into capturing design rationale, they may be rewarded with a resource that will not only enhance their design decision making process and provide a more structured approach to their product design, but when collated over a number of design seasons, it can further be used to support their overall strategic planning.

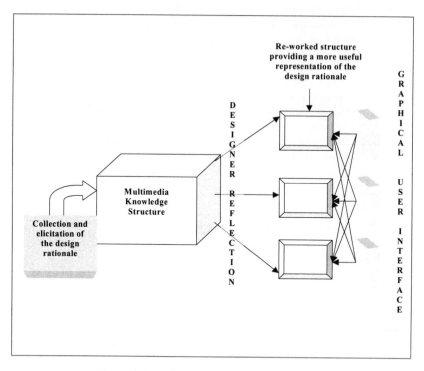

Figure 10 Research conducted and planned future work

REFERENCES

Kunz, W., and Rittel, H.W.J., 1970, *Issues as Elements of Information Systems*, Working Paper No. 131, (Berkley: University of California).

Moran, T. P., and Carroll, J. M., (Eds.), 1996, *Design Rationale – Concepts, Techniques and Use*. (New Jersey, USA: Lawrence Erlbaum Associates).

Phillips, S., 1997, Multimedia Fabric Libraries for the Textile and Clothing Industry. In *Proceedings of 78th World Conference of the Textile Institute,*. Thessaloniki, Greece, May 1997. (The Textile Institute, Manchester, UK), Vol. 1, pp.397-431.

Designing over Networks: a review and example of using internet collaboration and communication tools in design

P.A. Rodgers and A.P. Huxor

Design has been described as one of the most demanding tasks faced by human beings, and also as an activity which frequently involves specialists from many areas, such as business, management, manufacturing, and marketing (Coyne *et al.,* 1990). A characteristic of contemporary design practice is the high number of trade-offs and decisions that have to be made between one specialist and another during the design process.

Recent social, market, and technological developments have led to fundamental changes in the activity of design, including the way these multidisciplinary design teams now function. These changes have been brought about as a result of many factors, including the increase in the life expectancy of the population, the limited economic life of many products, the pressures on designers and manufacturers to reduce the development time for products, the need to improve product quality, and last but not least the need for improved communication between designers, engineers, manufacturers, and other relevant parties. A particular issue is the emergence of the virtual design team (Line and Syvertsen, 1990) brought together to exploit market opportunities quickly and flexibly.

Recent advances in communication technologies which strive to support 'better' collaborative design work aim to improve the chances of successful new product design and development by facilitating more effective and efficient collaboration between members of design teams. Many Internet collaboration and communication tools now exist to support designers including shared product data storage, audio and video conferencing, real-time chat, threaded discussion lists, and so on.

This paper will review and compare the current state of Internet collaboration and communication tools for design and describe, by working through a design example, how these tools might be used to facilitate greater success in design practice. The example will show how a number of product design specification (PDS) elements (Pugh, 1991) can be taken into account within this collaborative approach. That is, how the collaboration and communication tools can aid designers work through issues commonly addressed throughout the design process, such as human factors issues, manufacturing and production issues, logistical issues, environmental issues, and so on.

INTRODUCTION

With the growth in global engineering markets, the success of new product design and development is becoming more dependent upon effective and reliable communication and collaboration between members of the design team. One of the key factors in the globally distributed marketplace of concurrent engineering (Nevins and Whitney, 1989) and continuous improvement is knowledge and information. Thus, knowledge and information has become a crucial factor and key resource within current flexible, distributed working practices (Court *et al.,* 1997). Today, the exchange of information is achieved via a wide variety of communication and collaboration media, including email, face-to-face, telephone, video conferencing, formal documents, etc. (Harmer, 1996). Recent studies of the knowledge and information requirements of designers (Kuffner and Ullman, 1991; March and Trott, 1988), and the media utilised in accessing this information illustrate that designers tend to favour the highest level of information richness (Daft and Lengel, 1984) such as face-to-face, telephone, and email during decision making activities in design (Court, 1995; Marsh, 1997; Rodgers 1997).

The remainder of this paper comprises three main sections. The next part of the paper, Section 2, contains a brief review of existing Internet-based collaboration and communication tools and techniques. This section is not intended to cover every tool and technique, but rather to provide a current picture of what is being used on a day-to-day basis by designers. Section 3 describes the main part of this paper. This section details the collaboration and communication during an on-going collaborative design project between globally distributed design team members using Basic Support for Co-operative Work (BSCW), a Internet-based collaboration and communication tool. This section charts the progress that has been made, to date, on a "real world" design problem example. The paper concludes with a discussion on the suitability of the BSCW tool for globally distributed design collaboration and communication.

REVIEW OF INTERNET COLLABORATION AND COMMUNICATION TOOLS

Current flexible approaches in collaborative and distributed design work are being made possible by new and emerging collaboration and communication technologies, particularly using the Internet. At present, designers use the Internet as a means for collaborating and communicating with others, and for sending and receiving design information. Methods used to communicate and collaborate with others include email, electronic list-servers and news groups, and Internet-based communication and collaboration tools. For example, designers frequently submit email queries to list-servers[1] for general design assistance, and browse news groups[2] for more specific information (usually CAD software related). More recently, there has been a significant rise in the amount of communication and collaboration provision over the

---

[1] design-research@mailbase.ac.uk; engineering-design@mailbase.ac.uk; engineering-cace@mailbase.ac.uk
[2] alt.cad; alt.cad.autocad; comp.cad.pro-engineer

Internet which facilitates "real time" chat, shared white-boards, shared drawing and painting boards, and video and audio conferencing. For example:

- Globe Java Chat[3];
- Multi-user White Board[4];
- Pixel Toy[5];
- WebDraw[6].

The Internet currently provides an effective and efficient mechanism which enables the bringing together of designers, from very different geographic locations, to work on collaborative projects. The next section of the paper will describe the results of an on-going collaborative design project wherein designers from four continents have been brought together using some of the latest Internet communication and collaboration tools to work on a design problem.

COLLABORATIVE DESIGN PROJECT

The background to this project was when an email was submitted to the Industrial Design Forum (IDFORUM@YORKU.CA) suggesting an *on-line mailing-list ID workshop*. The idea being that once the designers involved in the project had identified and established a suitable design brief, to which potentially everyone subscribing to the ID list would have an input, on-line presentations including background research, concept sketches, models, renderings, technical drawings, and so on could be posted to the on-line workshop. The project would then run its course during which discussions and decisions would be made on the development of the project resulting in a truly global solution.

This idea happened to be very close to a project proposal, which was under development at the same time by the authors, in which globally distributed designers could be brought together to work on the same design project over the Internet. A truly global collaboration commenced shortly afterwards, and BSCW was proposed as a suitable medium for the collaborative design project.

BSCW (Basic Support for Co-operative Work) developed by GMD FIT (http://orgwis.gmd.de/), a form of CSCW (Computer Supported Co-operative Work), supports group work over the Internet by offering 'shared workspaces'. These workspaces offer many possibilities including the uploading and downloading of documents, the addition of group members and web sites, and the distribution of information within groups or teams (Figure 1). BSCW was quickly and enthusiastically agreed as a simple and effective medium for the collaborative project by all members of the design team.

---

[3] http://www.theglobe.com/chat/rapturecafe/entry/

[4] http://power.eccosys.com/WB/

[5] http://www.talk.com/pixeltoy/index.html

[6] http://www.microsurf.com/WebDraw/

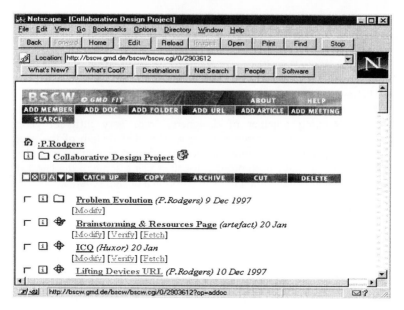

Figure 1 BSCW workspace

Collaborative design team structure

The importance of effective and efficient co-operation and communication of knowledge and information between members of design teams and the role this plays in the overall success of the product design and development process is well acknowledged (Bush and Frohman, 1991; Meerkamm, 1997).

During the early discussions of the collaborative design project three to four individuals expressed a keen interest, whilst many other IDFORUM members expressed real fears and reservations about such a project. The majority of these fears were related to issues such as Intellectual Property Rights (IPR), credit for design ideas, and payment. These issues did not dissuade many individuals from expressing an interest in the project, however, and the BSCW collaborative design team, at present, now comprises twelve designers from various geographical locations throughout the world (Figure 2). Four designers (RL, EO, JV, SM) are located in the USA, three in Australia (CH, CF, PH), two in Canada (MD, LG), two in the UK (PR, AH), and one in India (AL).

The design team consists of designers with a wide variety of backgrounds and interests including medical and rehabilitation equipment design, packaging and graphic design, toy design, architecture and exhibition design. Particular specialist areas covered by team members include manufacturing methods and production techniques, marketing, and psychology. The project members are based in a number of employment scenarios including government organisations, consultancy groups, freelance work, and education.

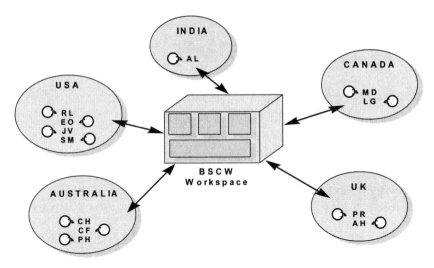

Figure 2 BSCW collaborative design team

Design problem

A design problem was identified and submitted to the collaborative workspace approximately one week after the initial discussions began. The design problem originated from one of the design team members (EO), located in the USA, after she observed and discovered problems 'first hand' carrying heavy loads up stairs.

The initial design problem statement set, and agreed by all members involved in the project as a non-trivial problem worthy of investigation, reads: "There is a need for low cost, low tech-high function, good looking elevation devices that can get people and things up and down interior or exterior stairs. This is especially problematic in existing buildings that have a variety of conditions. For example, if you're an artist with a studio on the 4th floor of a walk-up, hauling 100 pounds of clay is tough. If you have a broken leg, it's another problem that might have a similar solution".

From this initial problem statement, the design project progressed through many stages. The iterative stages and some of the tasks undertaken in the collaborative design process, to date, are outlined in Figure 3. Here, the shaded boxes define the stages the project has progressed through (to date) whilst the non-shaded boxes are stages yet to be reached. Further stages of the process such as manufacturing, marketing, product costing, and selling are purposely not included in the figure.

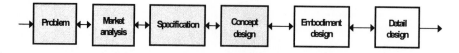

Figure 3 BSCW collaborative design process (to date)

Problem exploration

Problem exploration, sometimes referred to as clarification of the task [Pahl and Beitz], involves collecting information about the requirements that have to be fulfilled by the product, and also about the constraints and their relative importance. The end result of this task is an objectives or requirements list. The design problem here has been explored and developed further since the initial problem was set. For example, one designer (PH) suggested that the problem could be extended to include: "...carrying the load over very rough terrain...which may require some level of self adjustment for the changing conditions..."

Several other group members (SM, MD, PR, AL) explored the problem statement by developing and describing specific aspects of the problem statement, such as load, terrain anticipated, motion. This is illustrated below: "...load - volume, weight, centre of gravity, liquid, solid, sheets, blocks, pellets... ...terrain - pavement, cobblestones, ramps, stairs, rough, slippery, wet, rubbery... ...motion - forward, backward, sideways, crawling, jumping, rolling, floating...".

One of the other designers (EO) explored more specific areas of the problem by asking other group members to define certain parts of the problem statement: "...some thoughts about 'low cost' and 'low tech' - this is pretty subjective, and these figures assume some volume of production to achieve this lower cost. I think that anything under $1,000 for equipment that can be installed in a building or on a site and can carry stuff is 'low cost'. For anything that can carry a person, installed in a building or on a site, $5,000 is 'low cost'. For a portable device, it may have to be $500 or lower. But this has the potential to be something more...I am less sure about this. Regarding 'low tech' - I'd love some discussion about this. What are options for powering?...".

As the problem statement developed, conjectures thrown up for consideration by the design team members included concepts based on existing products that they either had first hand experience of using, or could recall from personal memory. This is interesting as it concurs with the broad view that designers rely heavily on their past knowledge and experiences during problem exploration, and also when proposing initial ideas that attempt to meet the problem (Duffy and Kerr, 1993; Goker and Birkhofer, 1995).

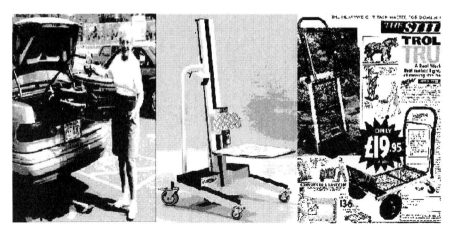

Figure 4 Examples of load carrier competitor products

Table 1 IBM patent search results for "Load Carrier" query

Patent Number	Patent Title	Inventor	Date of Patent
5697757	Counter-balanced load carrier	R.A. Lindsay, U.K.	16 Dec. 1997
5560525	Load carrier	H. Grohmann and Christiansson, Sweden	1 Oct. 1996
5524803	Load carrier	J.I. Arvidsson, Sweden	11 June 1996
5356260	Bucket carrying loader	M. Ikari and M. Fukuda, Japan	18 Oct. 1994
5138953	Load carrier suspended from rail	W. Horcher and S. Horcher, Germany	18 Aug. 1992

The analysis of design knowledge and information currently available, and various activities pertaining to the problem exploration stage, including a patent search was carried out by one of the designers (SM) to ensure that no possible reinvention would occur, and also to see if there were any immediate improvements that could be made to existing solutions. A very brief summary of the results from the patent search, using IBM's U.S. patent Database (http://patent.womplex.ibm.com /boolquery.html), is given in Table 1. Similarly, two other members of the design team (PR, CH) conducted an Internet search for information on competitor and like products. A small selection of the results of this market analysis are shown in Figure 4.

The main outcome from the problem exploration stage was a formal definition of what was required of the load carrying device. Requirements of the load carrying device were submitted to the shared workspace by several designers (MD, SM, EO, CH, PR, LG, PH, AL), and summarised in a objectives tree. The objectives tree for the load carrying device is shown in Figure 5.

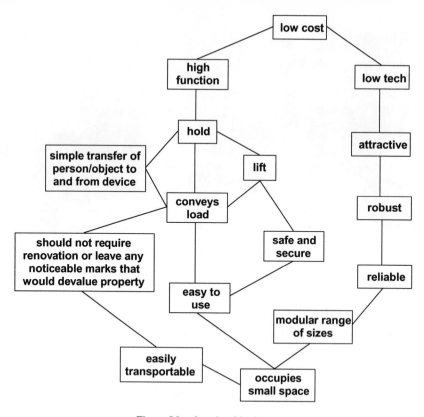

Figure 5 Load carrier objectives tree

Concept design

Concept design involves the generation of ideas usually represented in a suitable form, such as a design sketch, diagram, or scale model. This highly iterative stage involves the generation of ideas (synthesis), and the checking or evaluation of those ideas against the objectives or requirements (analysis). The objectives tree (Figure 5) was used as a basis for firstly generating and secondly evaluating the concept design ideas. The guidelines for brainstorming ideas, suggested by Cross (1994), were used during this activity. These are as follows:

- no criticism is allowed during brainstorming;
- a large quantity of ideas is wanted;
- seemingly crazy ideas are wanted;
- keep all ideas short and snappy;
- try to combine and improve upon the ideas of others.

The concept design ideas generated by the group members, to date, have taken three forms:

- concept design text-based descriptions;
- concept design sketches;
- 3-D computer models.

The reason for the disparity in formats is due to the fact that some members of the design team found difficulty in sending their concept sketches to the BSCW workspace, and relied instead upon sending textual descriptions of their ideas by email. A selection of early text-based concept ideas submitted by their designers are given below: "(MD)...ski lifts (chair lifts, gondolas), grain elevator, conveyer belt, catapults, slingshots etc... ...air systems that are used to move large machinery (several tons)... (LG)...waddling cartoon-type figure rocking side to side... ...walking horse where the back "leg" swings forward when the horse rocks forward - textured back feet grab the floor and stay there, while the front feet are not textured and slide the whole horse forward.... (AL)...saddle-like device that hangs of the stair banister and can be cranked... ...wicker basket seat slung under a wood log carried by two able persons... (CH)...three wheels arranged with their axes at the points of a triangle on each end of an axle... ...pedal powered stair climber... ...arm pumped ratchet device... ...donkey at the top or bottom of a hill driving an endless loop of rope to which you attach buckets of cargo...".

A selection of some of the early concept design sketches submitted by design team members are illustrated in Figure 6.

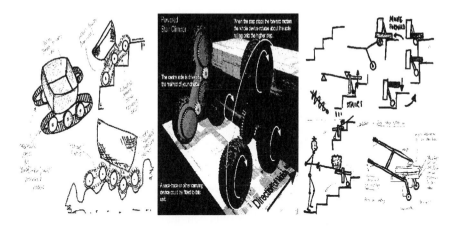

Figure 6 Load carrier concept sketches

Here, three concept sketches of differing levels of representation are shown. The concept on the left depicts a caterpillar track load carrier driven by a battery or electrical power source. The middle concept shows a powered stair climbing device (optional power source) which rotates about the axle allowing the device to 'flip' onto the higher step. The concept on the right of the three illustrates a simple 'wheelbarrow-style' load lifter with hydraulically-powered bars (optional) to assist lifting.

Collaborative design project summary

To date, the collaborative design project has successfully progressed through a number of design stages and activities. The activities undertaken include the identification and definition of a problem, problem exploration, market analysis, patent searches, and concept synthesis and analysis. However, the project is now currently at a crucial stage. Key issues such as determining the intended market for the product, the product retail cost target, methods of manufacture, and other details of the product specification have to be agreed.

To this juncture BSCW has facilitated effective collaboration and communication between the group members. However, it is clear that face-to-face meetings are now required, that will provide the appropriate level of communication 'richness' (Daft and Lengel, 1984), to resolve some of the more key issues mentioned above. This is obviously impossible, so an Internet-based communication tool which provides a high level of 'richness' is required. Currently the group members are investigating the use of ICQ, an Internet communication tool (http://www.mirabilis.com/icqme.html), as a method for enabling 'real time' chat to address some of the project issues which have arisen.

CONCLUSIONS AND DISCUSSION

To date, the most significant problem of using BSCW in this project has been the different hardware platforms and software applications used by individuals. Generally, however, the designers involved in the collaborative design project found BSCW a relatively straightforward and effective system to use. BSCW features utilised most, to date, include the uploading and downloading of documents and information. However, other functionality BSCW provides such as the addition of "articles" and "notes" to documents, document "versions", and document "descriptions" were hardly used. This is probably due to the fact that once the group members understood the basic features of BSCW they did not wish to explore the system any further.

BSCW's integration with email is one of the most practical and useful features that has been used to date in this project. Hundreds of notes and messages have been sent by all designers involved in the project. One significant disadvantage of email communication, however, is that the sender can never be certain whether the recipient has read the message. BSCW addresses this weakness by the "Readers of..." function. This function allows the user to see which members of the group have read the message or document that has been sent to the workspace and when it was read. Furthermore, BSCW's integration with the World Wide Web (WWW) allows group members to submit links from the BSCW shared workspace to large amounts of relevant on-line information.

Probably the greatest weaknesses of BSCW are that it does not facilitate "real time" or "face to face" communication, nor offer a facility for collaborative sketching / drawings in a shared "white board" setting - features crucial to successful design working practice.

REFERENCES

Bush, J.B. and Frohman, A.L., 1991, Communication in a 'Network' Organization. In *Organizational Dynamics*, Vol. 20, No. 2, pp. 23-36.

Court, A.W., 1995, *The Modelling and Classification of Information for Engineering Designers*, PhD Thesis, University of Bath.

Court, A.W., Culley, S.J. and McMahon, C.A., 1997, The Influence of Information Technology in New Product Development: Observations of an Empirical Study of the Access of Engineering Design Information. In *International Journal of Information Management*, Vol. 17, No. 5, pp. 359-375.

Coyne, R., Rosenman, M.A., Radford, A.D., Balachandran, M. and Gero, J.S., 1990, *Knowledge-Based Design Systems*, (Reading, M.A. Addison-Wesley).

Cross, N., 1994, *Engineering Design Methods: Strategies for Product Design*, (Chichester: John Wiley & Sons).

Daft, R.L. and Lengel, R.H., Information Richness: A New Approach to Managerial Behaviour and Organisation Design, 1984. In *Research in Organisational Behaviour*, **6**, pp. 191-233.

Duffy, A.H.B. and Kerr, S.M., 1993, Customised Perspectives of Past Designs from Automated Group rationalisation. In *Artificial Intelligence in Engineering*, No. 8, pp. 183-200.

Goker, M.H. and Birkhofer, H., 1995, Incorporating Experience in the Methodological Design Process, In *V. Hubka (Ed.), Proceedings of ICED '95, Praha, Czech Republic,* pp. 1455-1460.

Harmer, Q., 1996, *Design for Low-Volume Production*, PhD Thesis, University of Cambridge.

Kuffner, T.A. and Ullman, D.G., 1991, The Information Requests of Mechanical Design Engineers. In *Design Studies*, Vol. 12, No. 1, pp. 42-50.

Line, L. and Syvertsen, T. G., 1996, Engineering Teams: Strategy and Implementation, In *Turk, Z. (ed.) Construction on the Information Highway*. Electronic Proceedings, http://www.fagg.uni-lj/bled96/.

March, J. and Trott, F., 1988, Accessing of Information for Engineering Design: Databases for Engineers. In *Professional Engineering*, Vol. 1, No. 7, pp. 29-32.

Marsh, R.J., 1997, *The Capture and Utilisation of Experience in Engineering Design*, PhD Thesis, University of Cambridge.

Meerkamm, H., 1997, Information Management in Design Process - Problems and Approaches to Their Solution. In *Designers – The Key to Successful Product Development* (Eds.) E. Frankenberger, P. Badke-Schaub and H. Birkhofer. London, Springer Verlag Berlin, pp. 249-264.

Nevins, J.L. and Whitney, D., 1989, *Concurrent Design of Products and Processes: A Strategy for the Next Generation*, (USA: McGraw-Hill Publishing Co).

Pugh, S., 1991, *Total Design*, (Reading, M.A: Addison-Wesley).

Rodgers, P.A., 1997, *The Capture and Retrieval of Design Information: An Investigation of the Information Needs of British Telecom Designers*, Technical Report, CUED/C-EDC/TR58, Cambridge University, Engineering Design Centre.

The Long-Term Benefits of Investment in Product Design and Innovation

R. Roy with S. Potter and J.C.K.H. Riedel

This paper concerns a study, entitled Market Demands that Reward Investment in Design (MADRID), which aimed to identify: how the commercial returns from investments in design and product development vary with the types of market in which a firm operates; and the long-term benefits of investment in product design and innovation. MADRID built upon an earlier research project on the Commercial Impacts of Design (CID). CID involved a survey of design and product development projects in 221 SMEs that had received some government support for design. This paper gives some results of a longitudinal, follow-up survey of a sample of 42 CID firms and product development projects, 8–9 years after the original study. The survey produced valuable results at both company and product levels – this paper focuses on major 'company-level' findings.

The firms which had grown rapidly in turnover over the past five years operated in growing markets and had typically developed innovative or niche products, while the declining firms generally operated in static or declining markets in which they had many competitors. However, nearly a third of firms, including several of the fast-growers, were still making the products developed with assistance under the Support for Design programme, which had been first launched between 8 and 14 years ago.

The fast-growing firms employed a higher proportion of Research, Design and Development) RD&D staff; more often used external expertise for product development; introduced new products more frequently; and were more likely to employ modern product development practices, than the slow-growing or declining firms. Most of these differences were statistically significant.

There was also a statistically significant relationship between management attitudes and company growth. All the growing firms had managers with a positive attitude towards design and innovation and increased their investment in RD&D during the recession, while most managers in the declining firms had a narrow and limited understanding of the contribution of design and had reduced their investment.

These findings support those of previous research that business success and investment in design and product development are likely to be mutually reinforcing, while poor financial performance and a failure to invest can lead to a cycle of decline. This study also supports the conclusions of other work that investing in design and product development is likely to be a necessary, but not sufficient, condition for good business performance.

BACKGROUND

From 1987-90, the Design Innovation Group undertook a study of design and product development projects in 221 small and medium-size UK manufacturers that had received government funds to engage a design consultant to help with the development of new or improved products. The firms were sampled to be representative of UK manufacturing industry as a whole and the projects embraced a wide range of products and technologies, from electronic instruments and railway equipment to textiles and food packaging. This Commercial Impacts of Design (CID) study provided unique information on the commercial returns upon investing in professional design expertise at the product level. (Full details may be found in e.g., Potter *et al.,* 1991; Roy and Potter, 1993).

Since CID there have been several other attempts to measure the commercial benefits of investing in design and new product development in SMEs. These include a study by Groupe Bernard Juilhet (1995) which examined the extent that a sample of 500 French SMEs invested in industrial design and the costs and benefits of these investments at the firm level. Another study investigated the commercial performance of 38 products which had won a Dutch Good Industrial Design Award (Roerdinkholder, 1995). Both studies indicated that investing in industrial design confers commercial benefits for firms and for products. More recent research conducted by Gemser (1997) compared the business performance of matched samples, from two sectors, of 20 Dutch SMEs that routinely employed industrial designers with 20 that did not. She showed that furniture and medical / industrial instrument firms which regularly invested in industrial design performed better than those which did not. More generally a study by Sentance and Clarke (1997) provided evidence of a positive relationship between the level of design expenditure in different manufacturing industry sectors and their rate of output growth over ten years.

None of these studies, however, considered whether investment in design and product development is dependent on the market in which a firm operates or the long-term benefits of investing in product design and innovation. A new study, entitled Market Demands that Reward Investment in Design (MADRID), examined these questions. The MADRID project was divided into two phases:

- Phase 1 involved a record analysis of the data on selected projects from the original CID study in order to identify: a) which types of market(s) are most likely to produce the best commercial returns from investments in design and product development; b) the contribution of design and innovation to product competitiveness.
- Phase 2 involved a longitudinal, follow-up survey of a sample of 42 CID firms and product development projects, 8–9 years after the original study: a) to explore relationships between business success, the nature of the market in which the firm operates and its management of design; b) to identify long-term benefits of investments in design and product development.

The firms and projects for both Phases were sampled to be typical of small and medium-size UK manufacturers.

The full results of Phase 1 have appeared in earlier publications (e.g. Riedel, *et al.,* 1996; Roy and Riedel, 1997). This paper therefore focuses on Phase 2 of MADRID.

THE MADRID SURVEY

Methodology

We chose firms from the CID database which had conducted product, engineering or engineering / industrial design projects, ranging from furniture and textiles to vehicle components and electronic equipment.

Before a visit was arranged a telephone interview was carried out to establish the suitability of the firm and to identify a 'selected product' to be the focus of the face-to-face interview. Where possible we selected the original product surveyed at the time of CID. But if that product was not in current production, or had become peripheral to the firm's business, we identified a suitable successor product or range.

A semi-structured questionnaire was designed and piloted in two firms. This questionnaire was then administered in a further 40 interviews, conducted from late 1996 to mid. 1997, with senior managers, marketing or technical staff.

A framework was established to allow comparison of the data obtained in the MADRID interviews with some of that from the original CID survey. This data was subjected to both computer-based and manual analysis. Some results at both product and company levels have already been published (Roy, *et al.,* 1998a; 1998b). This paper provides additional detail on the *company-level* findings.

THE MADRID FIRMS AND PRODUCTS

Firm size

Most (90%) of the firms surveyed were SMEs with below 500 employees. 2% were 'micro' firms with <10 employees; 51% had 10-99 and 37% had 100-499 employees. Since the CID interviews, the number of people employed by the firms had generally declined. An exception was that the 'micro' firms at the time of CID had grown in employment.

What happened to the CID products?

The MADRID interviews sought to discover what happened to the original product/range surveyed at the time of CID.

A surprising number (29%) of firms were still making the products developed with assistance under the government Support for Design programme, and which had been first launched between 8 and 14 years ago. Although this apparent lack of new product development can be regarded as negative, in several cases this was not so. For example, the manufacturer of an innovative front-opening bath had built up a substantial niche market around a design that needed little further development (Figure 1). In another case the original product, a range of hospital furniture, had not altered in design, but its manufacture had been automated, thus boosting profits. At the same time new product ranges were introduced.

14% of firms still made the original product but had substantially modified or redesigned it technically and/or aesthetically. For example, an innovative wire-joining device, proved to be an excellent core design that remained in production while variants for new markets were developed (Figure 2).

36% had replaced the original product with a new product or range in response to market and/or technical change (Figures 3 and 4). In many cases this formed part of an expansion of the company's product portfolio.

10% of the original CID products (e.g., a range of bus shelters) remained available to special order even though successor products had been developed. This occurred where customers still used the original product and wanted the same design when replacements were needed.

There were a variety of other developments. In one case the original supported project (a spring-powered torch) had not been implemented at the time of CID, but had subsequently been developed as a joint venture and was just about to be launched when we revisited the firm. At the same time the firm continued, through two changes of ownership, to use its constant force springs in the manufacture of a very successful range of pedestrian barriers (Figure 5).

Figure 1 Front-opening domestic bath for elderly and disabled users, developed with assistance from Support for Design, has remained in production with only minor modifications since its launch in 1989.

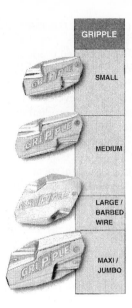

Figure 2 The Gripple patented wire joining device, the original design of which was created
with assistance from Support for Design, has been significantly improved since its introduction
in 1988. The design has been developed into a range of sizes for different applications
for example vine supports. It has also formed the basis of new products for new applications
such as a wire rope grip.

Figure 3 The Alpha 5+ hi-fi amplifier, launched in 1995, is one of a series of new and improved
designs that have evolved from the original Arcam Alpha of 1984 developed with assistance
under Support for Design. The latest model is the Alpha 7.

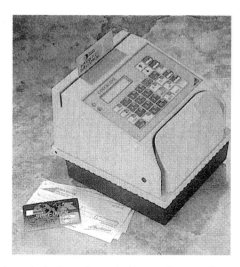

Figure 4 FASTtran electronic fund transfer terminal is one of a range of new products that replaced the supermarket cheque-writing machines originally developed with assistance under Support for Design.

Figure 5 The Mark 2 Tensabarrier range of pedestrian queue guidance systems has been developed from the original patented design introduced in 1978. A new Mark 3 design was introduced in 1997.

MARKETS, DESIGN MANAGEMENT AND COMPANY SUCCESS

In this section we examine relationships between the performance of the MADRID firms, the nature of the markets in which they operate and their management of design and product development.

Firm growth and market type

A number of measures of firms' performance were used in MADRID but the most satisfactory was turnover growth over the past five years, as this information was available from 39 of the 42 companies surveyed and is the performance criterion adopted by Hart and Service (1988) and Sentance and Clarke (1997) in their studies of design management and firms' success.

Significant break points were identified that divided the sample into almost equal quartiles, as follows:

- Very fast-growing   > 85% turnover growth
- Fast-growing   33 – 77% turnover growth
- Moderately growing   11 – 29% turnover growth
- Static or declining   0% to 59% turnover decline

However, a number of studies show that firm growth has to be considered in the context of the type of market in which the firm is competing. For example, Porter (1980) and Buzzell and Gale (1987) show that the ability of a firm to grow is related, among other things, to whether the market for a firm's products is growing, how many competitors the firm has in that market (competitive intensity), and the maturity of the products involved.

Thus, the types of market in which our sample firms were attempting to survive and grow range from 'very difficult' static or declining markets for mature products fought over by many competitors, to relatively 'easier' growing markets with few competitors (see Table 1).

Not surprisingly, we found a strong relationship between firms' performance and the type of market in which they sold their selected product or range. Table 1 shows that there are rapidly growing firms in three of the four market categories. However, 80% of the very fast-growing firms, and 66% of all growing firms, operated in growing markets and only 20% and 33% respectively were in static/declining markets. In addition 7 of the 10 fastest growing firms had developed an innovative or a specialist niche market product, often with few competitors.

For example, the fastest growing firm in the survey had produced a novel product, the wire-joining device mentioned above, for which it was continually finding new markets and applications and as a result had grown in turnover by 1400% and expanded from three to 40 staff in the past five years. Another very fast-growing firm produced a range of innovative electronic fund transfer terminals for which there was rapidly growing demand. It had grown from two rooms above a shop employing six people when we visited in 1989 to its own headquarters and 34 staff. A third very fast-growing firm had developed a range of flame-retardant fabrics, for a specialist contract market and had grown by 175% over the past five years and by over 1600% in the 8-9 years since the CID interview.

By contrast over 83% of the declining firms operated in static or declining markets and in all but one case they were fighting against several competitors. A typical example was a firm that made shoe repair machinery. The market had declined because nowadays fewer people have shoes repaired. In addition the firm had German, Dutch and Italian competitors. To compete the firm had to develop new designs, but given the declining market lacked the resources to do more than produce updated versions of a machine introduced 10 years ago.

Table 1 Company growth and type of market for the selected product / range

NATURE OF MARKET	Market Static or Declining in past 5 years	Market Growing in past 5 years

Competitive market	**1065 FLOWMETER**
	1025 TORQUE
	   WRENCHES
	
	1053 Bus shelter
	1073 Lorry trailer
	1242 Hi-fi loudspeaker
	
	1230 Water standpipe
	1032 Bicycle lock
	1040 Kitchenware
	1054 Bar stool
	1064 Car park barriers
	
	1010 Luggage
	1255 Cockpit light dimmer
	1072 Car sunroofs
	1085 Shoe repair equipment
	1050 Lorry cab
	1278 Analogue panel meters
	1043 Mirror panelling
	1030 Street furniture
	1332 Outdoor clothing
		1044 FUND TRANSFER
		   TERMINALS
		1038 HI-FI EQUIPMENT
		1293 ROPE HOLDING
		   SYSTEMS
		
		1306 Contract furniture
		1083 Car seat adjuster
		1058 Food cabinet
		1312 Wet/dry vacuum
		   cleaner
		1011 Rail switches/
		   crossings
		1029 Temperature
		   controller
		1080 Packaging
		   equipment
		
		1238 Anaesthetic machine
		1223 Litho plate processors
		1037 Ceramic tableware
		1224 Personal filing cabinet
		
		1256 Desk modules

Less competitive Market	*1047 Packaging machinery*	**1063 WIRE-JOINING**
		   DEVICE
		1005 TEXTILES
		1061 BATH
		1281 PEDESTRIAN
		   BARRIERS
		1077 WIND TURBINE
		
		1217 Valve tester

Key: **VERY FAST-GROWING FIRMS**
          1281 = Fast-growing firms
          1025 = Moderately-growing firms
          1030 = Static/Declining firms

Note: Firms are identified by code number/selected product and are listed within each box in rank order of turnover growth. Two additional declining firms have been added to this matrix for which the exact decline is not known.

Overall, the relationship between firms' growth and market growth was highly statistically significant (Chi-Square p < 0.003).

It is notable that the nature of the market and the type of products made affected whether the firms had grown in turnover alone or in both size *and* turnover. In general the firms which achieved rapid turnover growth in competitive markets had done so by reducing employment, thus increasing productivity, while most firms operating in less competitive markets managed to grow in both size and turnover and hence displayed slow-growing or declining productivity.

Company performance and design management (employment of RD&D staff)

Other research (e.g., Sentance and Clarke, 1997) has indicated that successful manufacturers invest greater resources in Research, Design and Development than less successful ones. We examined whether the human resources employed in RD&D had any effect on company turnover growth.

Table 2 Company growth and employment of RD&D staff

Full-Time RD&D Staff as % All Employees	Very fast and Fast growing Firms	Moderately growing and Static /Declining firms
0 < 5%	7 (35%)	12 (66.7%)
5 – 20% +	13 (65%)	6 (33.3%)
TOTAL	20 (100%)	18 (100%)

Table 2 shows that there is a statistically significant relationship between firm growth and the proportion of full-time RD&D staff employed (Chi-Square p = 0.05). The fastest growing firms generally had a high proportion of staff involved in RD&D, the number depending on the sector. For example, the electronic fund transfer systems firm involved nine of its 34 staff in RD&D while the fire-retardant fabric manufacturer involved two of its 40 staff, both with relevant qualifications. Although not shown in the table, the slower growing and declining firms relied more on individuals doing RD&D as part of other tasks. The shoe repair machinery firm was typical here; the MD was responsible for design with assistance from the works manager. Neither had any technical or design qualification.

As well as employing more in-house design staff, the growing firms made greater use of outside expertise (e.g., design consultants) for product development. There is a statistically significant relationship between the use of external design inputs to RD&D and turnover growth (Chi-Square p = 0.006). All the fastest growing firms used external expertise, whereas two-thirds of the static and declining firms did not.

Development of new and improved products

Investment in new product development and improvement is often associated with business success (e.g., Hart and Service, 1988). In our sample we found a significant positive relationship between the frequency with which the firms introduced new products and their turnover growth (Table 3, Chi-Square p = 0.096).

Table 3 Company growth and new product introduction

Frequency of New Product Introduction	Very fast and Fast growing Firms	Moderately growing and Static /Declining firms
Annually or more	11 (68.8%)	4 (36.4%)
Less than annually	5 (31.2%)	7 (63.6%)
TOTAL	16 (100%)	11 (100%)

The fastest growing firms – such as the wire-joining device manufacturer, the fund transfer systems firm, the fire-retardant fabric supplier, the pedestrian barrier and the hi-fi manufacturer – had all continuously improved their products and developed a family of related products in the time since the CID survey. In contrast some of the static and declining firms based a major part of their business on the same, or slightly modified, versions of products developed 10-14 years ago with government support.

These findings were reflected in the firms' development of new and updated products during the recession. There was a positive relationship between increases in RD&D investment in the late 1980s/early 1990s and turnover growth (Table 4, Chi-Square p= 0.078). Thus, the MD of one very fast growing firm said "...it is vital to invest in design when times are bad, to sustain market growth through new and improved products."

Table 4 Company growth and investment in RD&D during the recession

Investment in RD&D of new/updated products	Very fast and Fast growing Firms	Moderately growing and Static /Declining firms
Reduced/no change	3 (33.3%)	8 (72.7%)
Increased	6 (66.7%)	3 (27.3%)
TOTAL	9 (100%)	11 (100%)

Management attitudes towards design and innovation

As the above quote suggests, company success is often associated with positive management attitudes towards investment and innovation. In our sample, we found a highly significant relationship between management attitudes and company growth (Table 5, Chi-Square p = 0.005).

All the growing firms had managers with a positive attitude towards the role of product design (and, where appropriate, innovation) and recognised their importance to the success of the firm now and in the future. Design and innovation are "absolutely vital" to the future, said a senior manager of the fast-expanding electronic fund transfer systems company because, for example, all its products will soon have to be designed to accept smart cards for payment. By contrast the managers in the declining firms predominantly had a limited and narrow understanding of the contribution of design to the success of the firm.

Table 5 Company growth and management attitudes

Management attitudes to design and innovation	Very fast and Fast growing Firms		Moderately growing and Static /Declining firms	
Very positive or positive	20	(100%)	12	(66.7%)
Limited understanding	0	(0%)	6	(33.3%)
TOTAL	20	(100%)	18	(100%)

Use of modern product development practices

Modern product development practices were common among these SMEs and almost universal among the growing firms. For example, one very successful firm has concurrent engineering teams for product development, who use CAD plus rapid prototyping facilities for design and physical modelling of products. However, nearly a third of the static and declining firms did not use such modern practices at all.

While there appeared to be no obvious relationship between company growth and the use of any single modern product development approach, there was a significant relationship between firm growth and the use of any of the modern product development practices considered together (Table 6, Chi-Square $p = 0.022$).

Table 6 Company growth and modern product development approaches

Product development Approach	Growing firms		Static/ Declining firms	
Design Reviews	4	(11.8%)	0	(0%)
CAD/CAM	17	(50.0%)	5	(38.5%)
Rapid prototyping	2	(5.9%)	1	(7.7%)
Concurrent Engineering	9	(26.4%)	3	(23.1%)
None	2	(5.9%)	4	(30.7%)
TOTAL	34	(100%)	13	(100%)

Note: Some firms responded in more than one category (total responses = 47)

CONCLUSIONS

Since the results of the Commercial Impacts of Design study appeared in the early 1990s, there have been several attempts to measure the commercial benefits of investing in design and new product development in SMEs. None of these studies, however, considered the long-term benefits of investing in product design and innovation. In our follow-up survey less than 20% of the firms contacted had gone out of business since the original CID study. This is a good record, given the severe UK recession of the early 1990s, and suggests that SMEs that had sufficient interest in product development to apply for government design support might perform better than more typical firms. Nevertheless, several firms, especially in the engineering and building products sectors, had contracted significantly in the recession and many had experienced severe financial problems and one or more changes of ownership.

Other findings of the MADRID survey indicated statistically significant relationships between business success and various measures of long-term investment in design and innovation. Thus, the firms which had grown rapidly in turnover over the past five years operated in growing markets and had typically developed innovative or niche products, while the declining firms generally operated in static or declining markets in which they had many competitors. The fast-growing firms employed a higher proportion of RD&D staff, more often used external expertise for product development, and introduced new products more frequently, than the slow-growing or declining firms. Fast-growing firms also increased their investment in RD&D during the recession, while slow growing and declining firms reduced or made no change in RD&D investment.

There was also a highly significant relationship between management attitudes and company growth. All the growing firms had managers with a positive attitude towards investment in product design and, where appropriate, technical innovation. By contrast the declining firms predominantly had a limited and narrow understanding of design and innovation and their relevance to the firm.

The above findings are in broad agreement with other research in this field e.g., Hart and Service (1988), Walsh *et al.,* (1992), Sentance and Clarke (1997). The MADRID survey also provides some encouraging evidence that some UK SMEs have moved beyond thinking in terms of financial returns on one-off design and product development projects and have made design and innovation an integral part of corporate strategy.

The MADRID project confirms once again that the relationship between investment in design and business performance is complex and interactive. As Gemser (1997) notes, "successful firms are more likely to have the resources to invest in design than those in financial difficulties". Managers in firms operating in growing markets had a more positive attitude to design and innovation than those whose firms were in declining markets. This suggests an interactive relationship – with market growth, positive attitudes and company growth reinforcing each other. In other words business success and investments in design and product development are likely to be mutually reinforcing, while poor financial performance and a failure to invest can lead to a cycle of decline.

Finally the analysis supports the conclusions of other work (e.g., Walsh *et al.,* 1992; Maffin et. Al., 1997) that investing in product design and innovation is likely to be a necessary, but not sufficient, condition for good business performance. Indeed the firms that managed to grow fastest in both turnover and employment were in general well managed all round, investing in training, marketing, manufacturing, quality, etc. as well as in continuous development of their product range.

REFERENCES

Buzzell, R.D. and Gale., B.T., 1987, *The PIMS Principles. Linking Strategy to Performance*, (New York: The Free Press).

Gemser, G., 1997, *Industrial Design for Competitiveness,* Paper for Second European Academy of Design Conference, Stockholm, Sweden, 23-25 April.

Hart, S.J. and Service, L.M., 1988, The effects of managerial attitudes to design on company performance. In *Journal of Marketing Management*, Vol.4, No.2, Winter, pp. 217–229.

Maffin, D. Twaites, A., Alderman, N., Braiden, P. and Hills, B., 1997, Managing the Product Development Process: combining best practice with company and project contexts. In *Technology Analysis and Strategic Management*, Vol.9, No.1, pp. 53-74.

Porter, M., 1980, *Competitive Strategy*, (New York: Free Press).

Potter, S., Roy, R., Capon, C.H., Bruce, M., Walsh, V.M. and Lewis, J., 1991, *The Benefits and Costs of Investment in Design.* Design Innovation Group, Report DIG-03, (Milton Keynes: The Open University and UMIST).

Riedel, J., Roy, R. and Potter, S. 1996, Investment in Design: A Market Analysis using the MADRID map. In *Proceedings 8th. International Forum on Design Management Research and Education*, Vol.2, (Boston: Design Management Institute).

Roerdinkholder, F.A. 1994/5, *Economische waarde van Goed Industrieel Ontwerp,* (Amsterdam: Netherlands Design Institute).

Roy, R. and Potter, S., 1993, The commercial impacts of investment in design. In *Design Studies*, Vol.14, No.2, April, pp. 171-193.

Roy, R. and Riedel, J., 1997, Design and Innovation in Successful Product Competition. In *Technovation*, Vol.17, No.10, October, pp. 537-548.

Roy, R., Riedel, J. and Potter, S., 1998a, *Market Demands that Reward Investment in Design (MADRID):* Final Report submitted to the Design Council, Design Innovation Group, (Milton Keynes: The Open University), January.

Roy, R., Riedel, J. and Potter, S., 1998b, Firms and Markets that Profit from Investment in Design and Product Development, Paper submitted to *The Design Journal*, February.

Sentance, A. and Clarke, J., 1997, *The Contribution of Design to the UK Economy*, (London: The Design Council), June.

Walsh, V., Roy, R., Bruce, M. and Potter, S., 1992, *Winning by Design: Technology, Product Design and International Competitiveness*, (Oxford: Blackwell).

ACKNOWLEDGEMENTS

This work was part of the Co-Partnership Programme of The Design Council, whose support is gratefully acknowledged. Dr Riedel is now at the Department of Engineering and Operations Management, University of Nottingham.

The authors wish to thank Mark T. Smith and Barry Dagger who helped with the interviews and all the individuals in the companies we visited without whose co-operation the research could not have been undertaken.

'Prototype Theory' and the Modelling of New Product Perception

John P. Shackleton and Kazuo Sugiyama

This paper uses the Japanese 'recreational vehicle' market as the basis for investigation into the way in which consumers differentiate between similar products, and how products are categorised into product groups by consumers, based on visual associations. The work demonstrates the inadequacy of 'classical' categorisation theory (i.e., that based on definitive sets of 'necessary and sufficient' criteria), as a basis for understanding these phenomena, and instead proposes the use of a 'prototype theory' model, taken from the field of cognitive science. This model postulates that judgement of group membership is achieved by assessment of similarity to the demonstrated 'focal centres' (or 'prototypes'). That is, a product will be judged to be a category member if a sufficient degree of similarity to the category 'prototype' is achieved. Demonstrating the presence of the fundamental elements of the model shows the applicability of Prototype Theory in product design domain. Firstly, the existence of 'similarity scales' is established. These similarity scales are attribute-based constructs by which observers determine the similarity of observed examples. Subsequently, a number of cognitive 'focal centres' within the product range are demonstrated. These focal centres appear to represent the category 'prototypes', and can be defined by a combination of quantified attribute values. These values appear to be central to the concept associated with a particular product group. It is also shown, however, that despite these clearly perceived bases for the groups, there are no 'necessary and sufficient' criteria by which the boundaries of the groups can be defined. Group membership is attribute dependent, but there can be a degree of trade-off in the attributes by which membership is achieved. Finally, the study demonstrates a major prototype effect, that of 'typicality'. Typicality is a measure of how representative a particular example is of its group, and it is shown that this may be successfully estimated from the attributes of form. It was also found that the 'typicality' of new vehicles tends to be lower than less recent examples. This suggests that vehicles come to be judged as more typical of their type as time progresses, and in this way the 'prototype' of a group may change. The results thus support the possible application of a Prototype Theory model to product planning and suggest that the optimum evolutionary development step may be achieved by controlling the typicality function.

INTRODUCTION

In the opening chapter of his book 'Women, Fire, and Dangerous Things', Lakoff (1987) has this to say of categorisation: "Categorisation is not a matter to be taken lightly. There is nothing more basic than categorisation to our thought, perception, action, and speech. ..... Whenever we reason about <u>kinds</u> of things - chairs, nations, illnesses, emotions, any kind of thing at all - we are employing categories".

Research has also shown that consumers' choice of products is profoundly affected by how they categorise them. This can function in a number of ways, but on the simplest level, if a product falls within the category of objects perceived by a consumer to meet their needs (both functional and symbolic) then the consumer will give consideration to the possibility of purchase. Conversely, if the product is perceived to fall outside that category then it will not be considered. However, it has been shown that most consumers also seek to express a degree of individuality in their product choices (Foxall, 1994). Thus, whilst the extent of individualistic expression sought can depend both on the attitude of the customer and on the type of product under consideration, there exists contradictory needs to express both belonging and individualism.

As a result, the way consumers differentiate between products should be of concern to designers, as it relates two important but conflicting requirements for successful product design. Firstly, it is essential that consumers can differentiate between competing products, and yet secondly, the semantics of a product must still identify it as a member of its correct functional group. Moreover, a user's first interaction with a product is usually visual, and first impressions can be very strong. If a product is not initially categorised as potentially meeting the consumer's needs it is likely to be rejected at this early stage and no further information will be sought. Thus, even though a product may be functionally ideal, it may fail by conveying the wrong associations. Designers should be careful, therefore, that in seeking visual differentiation they do not shift the identity of a product outside that of the intended product group. Similarly, to gain consumer acceptance of a very innovative product it may be necessary to signal its function by semantically relating it to existing concepts.

Thus, if it is possible to determine the way in which consumers categorise products, designers may find guidance as to the 'envelope' within which designerly manipulation of form can be successfully conducted. Some work on categorisation by consumers has been conducted in the context of consumer durables (Urban, *et al.,* 1993) but this has focused largely on the marketing and advertising rather than on design. What is of more use in the design process is a knowledge of how attributes are perceived by consumers, and how they affect the categorisation of the product in question.

The authors have conducted a number of studies on the possible application of Prototype Theory to this problem, and it is the intention in this paper to present an overview of the approach and the findings to date.

Origins of the work

An initial study by the authors (Shackleton and Sugiyama, 1996) on the recreational vehicle (RV) market in Japan, showed that this product group appeared to have evolved gradually out of its 4WD predecessor, the jeep, by a series of changes in both functional and styling attributes. However, classical categorisation theories require a fixed set of 'necessary and sufficient' criteria, and thus appear unable to accommodate such gradual emergence of new product groups. Within the field of cognitive science, however, 'Prototype Theory' (Rosche, 1975; Rosche and Mervis, 1975) has been successfully applied to a range of semantic domains, and its application to product design seems promising.

Prototype Theory

Stated simply, Prototype Theory postulates that classification is achieved by comparison to some category ideal, or 'prototype', and it has been suggested that this prototype is derived as some kind of 'average' of its members (although not in any strict mathematical sense). If this is so, then new members on the boundary of a category will not only be identifiable with the category, but, once included in the category, will subsequently affect the category norm. Assessment of later category members would then be made against this new norm. Furthermore, in terms of product design, if we imagine that items positioned close to the norm are perceived by consumers to be rather conservative, dull, or old-fashioned, and that designers consistently seek distinctiveness around the boundaries of the category, then the category norm will slowly but constantly shift. In this way, prototype theory not only accommodates stylistic evolution, but also predicts it.

Hampton (1993) outlines four elements of the standard 'prototype model': 'attributes' (and their associated values); a 'similarity scale', against which similarities are judged, based on a set of weighted attribute values; 'criteria', a level of similarity to a category norm which must be met by category members; and 'typicality', a judgement of how representative a particular member is of its category. This 'typicality' function should be monotonically related to the similarity of the particular member to the category 'prototype'.

This framework appears to match well with the product design activity; the 'attributes' are the attributes of design which are manipulated by the designer, and 'typicality' is the dependent function which the designer is seeking to control.

The designer, however, must also ensure that the relevant 'criteria' are met. In designerly terms, we may view 'originality' as the inverse of 'typicality', and thus the designer is seeking to maximise originality (i.e., minimise typicality) without moving outside the boundaries (or 'criteria') of the product type as perceived by the target consumer group.

The research work was thus aimed at demonstrating the presence of these elements, and hence the applicability of the prototype theory model. Attributes were taken as present *a priori* and the following sections take each of the three remaining elements in turn and demonstrate its presence in the domain of product design.

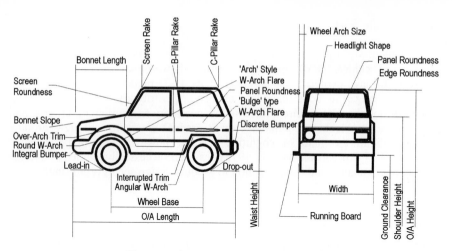

Figure 1 Vehicle attributes used in the analysis

EXPERIMENTS AND ANALYSES

Similarity scales

The first stage was to show that perceptions of similarity between products could be meaningfully related to their shared attributes, and the basis of the approach was a regression of perceived differences between vehicles against their differences in form.

Attribute data from 146 sample vehicles was collected, using a total of 34 categorical variables, (see figure 1, Table 1). This data was analysed using Non-linear Principal Components Analysis, a multivariate statistical techniques for handling categorical data, analogous to normal linear principal components analysis, (Gifi, 1990).

The effect of computing principal components is to remove any covariance from the data, and thus provide a distribution of sample points on a set of uncorrelated orthogonal 'attribute dimensions'. A rigid rotation using the Varimax criterion was used to obtain the 'most parsimonious' factors.

Perceived difference levels between vehicles were obtained from a paired comparison study, using a subset of 37 samples. Pairs of photographs were presented to subjects who were asked to make judgements of similarity on a seven point scale. Judgements from 120 subjects were normalised and averaged.

Absolute differences in object scores on each of the derived 'attribute dimensions' were calculated for all possible pairs of the 37 samples, and these differences were used as the independent variables in a stepwise linear regression against the similarity judgements from the paired comparison test. The aim of this was to find attribute dimensions that were strongly correlated with the judgements of similarity. Significant correlations were found with four dimensions, and inspection of the attributes that loaded on these allowed the nature of the dimensions to be identified.

Table 1 Attribute variables, categories and measurement levels

Attribute Variable	Variable Description	No. of Cats.	Category Description	Measurement Level
BMPRSTYL	Bumper Style	2	Discrete / Integral	Single Nominal
BNNTLEN	Bonnet Length	3	Short / Med. / Long	Ordinal
BNNTSLOP	Bonnet Slope	4	Level / Slight / Moderate / Very	Ordinal
BPILRAK	B-Pillar Rake	3	None / Moderate / Very	Ordinal
BPILTKN	B-Pillar Thickness	4	Thin / Slight / Moderate / Very	Ordinal
CDPILRAK	C/D-Pillar Rake	3	None / Moderate / Very	Ordinal
CDPILTHI	C/D-Pillar Thickness	3	Thin / Moderate / Thick	Ordinal
DOORS	Doors	4	Count	Single Nominal
DROPOUT	Drop-Out Angle	4	Large / Med. / Shallow / V.Shallow	Ordinal
EDGERNDN	Edge Roundness	3	Flat / Moderate / Rounded	Ordinal
GRDCLEAR	Ground Clearance	3	Low / Moderate / High	Ordinal
HEIGHT	Overall Height	3	Low / Med. / High	Ordinal
HLSHAP	Headlight Shape	3	Round / Rectangular / Wrap-round	Single Nominal
LEADIN	Lead-In Angle	3	Large / Med. / Shallow	Ordinal
LENGTH	Overall Length	3	Short / Moderate / Long	Ordinal
LHRATIO	Length/Height Ratio	3	Short / Moderate / Long	Ordinal
PNLRNDN	Panel Roundness	3	Angular / Moderate / Rounded	Ordinal
ROOFLINE	Roofline Roundness	3	Angular / Moderate / Rounded	Ordinal
RUNBOARD	Running Board	2	Absent / Present	Single Nominal
SCRNRAKE	Screen Rake	3	Upright / Raked / Very Raked	Ordinal
SCRNRNDS	Screen Roundness	3	Flat / Moderate / Rounded	Ordinal
SHLDHGHT	Shoulder Height	3	Low / Med. / High	Ordinal
SHLDLINE	Shoulder Line	2	Level / Stepped	Single Nominal
SOFTTOP	Soft-Top (Canvas)	2	Absent / Present	Single Nominal
TRIMSTYL	Side Trim Style	3	None / Interupted / Over-Arch	Single Nominal
WASHAPF	Front Wheel Arch Shape	3	Angular / Moderate / Rounded	Ordinal
WASHAPR	Rear Wheel Arch Shape	3	Angular / Moderate / Rounded	Ordinal
WASIZEF	Front Wheel Arch Flare Size	3	None / Small / Large	Ordinal
WASIZER	Rear Wheel Arch Flare Size	3	None / Small / Large	Ordinal
WASTYLF	Front Wheel Arch Flare Style	3	None / Arch Style / Bulge	Single Nominal
WASTYLR	Rear Wheel Arch Flare Style	3	None / Arch Style / Bulge	Single Nominal
WHEELBAS	Wheel Base Length	3	Short / Med. / Long	Ordinal
WIDTH	Overall Width	3	Narrow / Moderate / Wide	Ordinal
WSTHGHT	Waist Line Height	4	None / Low / Med. / High	Single Nominal

Table 3 shows the component loadings obtained from the Non-linear Principal Component Analysis, and the results of the stepwise regression analysis are shown in table 2.

Table 2 Regression coefficients for the selected factors

Factor	Factor Identity	B Coeff.	Std. Err(B)	Beta Coeff.	Sqrt. Beta	T Stat	Sig. of T
x1	Roundness	0.836	0.049	0.597	0.772	16.88	<0.0001
x2	Proportion	0.537	0.049	0.355	0.596	11.01	<0.0001
x5	Soft-top	0.214	0.053	0.130	0.360	4.04	0.0001
x3	Wheel Arch Emphasis	0.114	0.032	0.116	0.340	3.57	0.0004
(Constant)	-	3.293	-	-	-	14.95	<0.0001

Table 3 Component loadings from the Non-linear Principal Component Analysis

Attribute Variable	Rotated Factors									
	x1	x2	x3	x4	x5	x6	x7	x8	x9	x10
BMPRSTYL	**-0.787**	-0.241	0.151	0.198	-0.025	-0.176	0.060	-0.031	0.054	-0.041
BNNTLEN	**0.741**	0.255	-0.227	-0.074	0.253	0.282	0.118	-0.153	-0.143	-0.064
BNNTSLOP	**-0.788**	-0.142	0.007	0.157	0.104	-0.230	-0.176	0.070	-0.171	-0.110
BPILRAK	0.016	0.072	0.423	**-0.668**	0.356	-0.063	-0.129	0.117	-0.164	-0.175
BPILTKN	0.070	0.209	0.465	**-0.599**	0.010	-0.249	-0.089	0.125	-0.261	-0.036
CDPILRAK	-0.554	**0.537**	0.258	-0.092	-0.229	-0.055	0.098	-0.009	-0.046	0.177
CDPILTKN	-0.488	0.327	-0.301	-0.108	-0.149	0.229	0.021	0.333	0.180	-0.204
DOORS	-0.436	-0.071	0.025	0.246	0.052	0.087	-0.340	0.224	-0.554	0.085
EDGERNDN	**-0.638**	0.333	0.047	-0.154	0.431	0.194	0.290	-0.022	-0.062	0.091
GRNCLEAR	0.147	-0.099	0.159	-0.356	0.005	0.426	-0.382	-0.563	-0.056	-0.174
HEIGHT	0.566	0.344	-0.245	0.384	0.036	-0.040	0.085	-0.305	-0.197	0.213
HLSHAP	-0.052	-0.362	-0.249	0.326	0.243	0.226	0.113	-0.267	-0.409	-0.394
LEADIN	**-0.742**	0.435	0.168	-0.038	-0.230	-0.204	-0.026	-0.065	-0.058	0.029
LENGTH	-0.280	**0.842**	-0.242	-0.049	0.116	-0.104	0.002	-0.143	0.060	-0.033
LHRATIO	-0.298	**0.587**	-0.491	0.073	0.131	-0.258	0.106	-0.016	0.186	-0.178
PNLRNDN	**-0.854**	0.221	0.036	0.106	0.124	-0.017	-0.054	-0.161	-0.073	-0.120
ROOFLINE	**-0.764**	0.235	-0.198	0.149	0.223	-0.008	-0.019	-0.051	-0.131	-0.145
RUNBOARD	0.069	0.443	-0.031	0.190	0.254	0.454	-0.055	0.395	-0.116	0.034
RUNOUT	-0.461	**0.673**	0.142	-0.033	-0.189	-0.234	-0.020	-0.162	0.002	0.070
SCRNRAKE	**-0.672**	0.517	0.137	-0.245	0.028	0.007	0.002	-0.074	-0.004	0.140
SCRNRNDS	**-0.645**	0.366	0.072	-0.083	0.341	0.039	-0.024	-0.101	0.215	0.097
SHLDHGHT	0.404	0.412	-0.153	-0.370	0.308	-0.043	0.214	0.181	-0.095	-0.301
SHLDLINE	-0.174	0.351	**-0.517**	0.071	-0.175	0.134	-0.397	-0.017	-0.091	-0.188
SOFTTOP	0.302	-0.490	0.330	-0.003	**0.553**	-0.175	-0.004	-0.257	-0.064	0.072
TRIMSTYL	-0.410	0.421	0.236	-0.056	-0.235	0.230	0.338	0.000	-0.264	0.276
WASHAPF	**-0.662**	-0.367	-0.004	-0.261	-0.017	0.385	0.200	-0.090	0.076	-0.118
WASHAPR	-0.392	**-0.633**	-0.057	-0.035	0.282	0.321	0.322	-0.057	0.033	0.001
WASIZEF	0.127	0.390	**0.639**	0.535	0.174	0.008	0.123	0.053	-0.010	-0.161
WASIZER	-0.176	**0.592**	0.614	0.130	-0.197	0.287	-0.070	-0.046	0.100	-0.118
WASTYLF	0.048	0.413	**0.692**	0.422	0.215	-0.071	0.038	0.059	0.076	-0.197
WASTYLR	-0.213	**0.564**	0.644	0.077	-0.096	0.206	-0.134	-0.042	0.192	-0.183
WHEELBAS	-0.308	**0.810**	-0.246	-0.036	0.156	-0.132	-0.030	-0.103	-0.002	-0.059
WIDTH	-0.166	**0.790**	0.060	-0.089	0.116	0.241	0.093	-0.113	-0.221	0.182
WSTHGHT	0.154	-0.049	0.084	-0.094	**-0.501**	-0.123	0.447	-0.166	-0.380	-0.470

Criteria

The next stage of the research consisted of two analyses; the first aimed at demonstrating the presence of some underlying perceived groups amongst the samples, and thus, by implication, some perceived criteria for them; and the second aimed at demonstrating the lack of any clear 'necessary and sufficient' set of attributes to define these criteria.

Data for the first analysis was collected by a 'free grouping' exercise. Forty subjects were asked to sort photographs of sixty selected vehicles into groups that they considered similar. No restrictions were placed on the number of groups formed or the number of vehicles in each group, although subjects were encouraged to avoid single-member groups as far as possible. The collected data thus formed a matrix of samples by subjects, where the entry in each cell was a nominal group label. This data was examined by the use of Homogeneity Analysis (Gifi, 1990), a technique similar to Non-linear Principal Component Analysis but for purely nominal data.

The group labels in the data are valid only within-subject (i.e., group 1 for subject A is not the same as group 1 for subject B), and this is taken into account by the way the Homogeneity Analysis treats the data. As applied here, the algorithm in effect searches for pairs of samples that are consistently placed together by the test subjects. The more consistently any pair of samples is placed in the same group, the closer their position in the output space. The resulting distribution thus makes any commonly perceived groups evident. The output obtained is shown in figure 2. As the distribution in three dimensions is a little difficult to visualise, a schematic representation is also shown.

The second analysis also used the 'free grouping' data, this time together with the attribute data used previously. Discriminatory attributes used by subjects when classifying the vehicles were sought by using Non-linear Canonical Correlation Analysis (Gifi, 1990). Again a technique for multivariate categorical data, Non-linear Canonical Correlation Analysis determines the similarity between two or more data sets (rather than the variables in them). In this case, the objective was to determine how well the attribute data set could account for the distributions within the perceptual grouping data. The results comprise a sample distribution, similar to that obtained in from the Homogeneity Analysis, and component vectors descriptive of the output space. These are shown in figure 3.

Typicality

The objective of this final stage of the work was to demonstrate a method of estimating typicalities from the vehicle attributes, and to show that these estimates correlate with direct judgements of actual typicalities.

Logistic regression was used to estimate the typicalities. Product group memberships for the sample vehicles were established by using a cluster analysis on the sample distribution from the Homogeneity Analysis, and these memberships were used as the binary dependent variables.

The object scores provided the independent variables for the samples on the four previously established perceptual dimensions. However, as interpreting a continuous scale in terms of a single dichotomous variable would seem inappropriate, the original 'soft-top' binary attribute was substituted for this factor $x5$, and encoded as a dummy variable. The other three scales used, which had been previously determined to represent 'roundness', 'proportion', and 'wheel-arch emphasis', could all be satisfactorily interpreted as continuous scales.

Logistic regression constrains the predicted value of the dependent variable to be between zero and one, and these are usually viewed as probabilities. Here, however, the predicted values were interpreted as a measure of typicality, and there is some precedence for this (Feger and Boeck, 1993). In this way the logistic function makes it possible to evaluate the attributes of a sample and directly estimate the typicality of the sample within a particular category.

Regressions were conducted for each group membership indicator variable and the regression models obtained were used to predict typicalities and classifications for the other vehicles in the database.

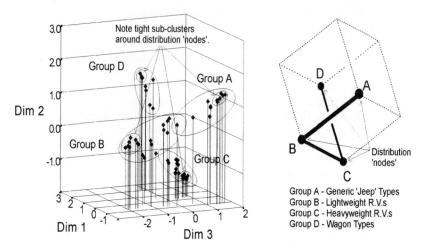

Figure 2 Output from the Homogeneity Analysis of perceived groupings

DISCUSSION

Inspection of the component loadings in table 2 suggests the first dimension is concerned with 'roundness'. The attributes which load highly on this axis are panel roundness, edge roundness, screen roundness and roofline roundness. Integral bumper styling also loads with the more rounded category values, suggesting it contributes to the overall roundness impression. Additionally, both short and sloping bonnets similarly load with the roundness factors, and perhaps these are perceived as part of the 'roundness' in the widest possible sense, i.e., as the overall shape approaches some kind of 'one-box' ellipsoid. Nonetheless, as the short sloping bonnet is related to the use of a transverse engine layout, from the design point of view it would be better if the bonnet attributes could be separated from the other attributes on this axis, but unfortunately, as the widespread introduction of this drive-train configuration is coincident with the general trend to rounder vehicles it does not seem possible to do so.

Attributes loading highly on the second factor comprise length, length/height ratio, width, and wheelbase. The run-out angle and to some extent the lead-in angle also load highly, and as these are largely dependent on the front and rear overhang lengths, the dimension appears to relate to aspects of proportion.

The third axis is loaded highly by a number of wheel arch attributes; front and rear wheel arch size, and front and rear wheel arch style. Thus this factor appears to be mainly concerned with the visual impact of the wheel arch areas.

The highest loading attribute on factor x5 is the presence (or absence) of a soft-top. The lack of any waistline on many open top jeep-type vehicles would appear to explain the covariance with the only other significant attribute on this axis, waistline-height.

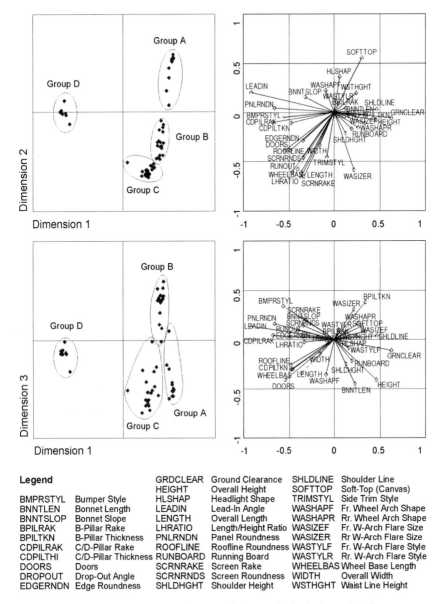

Figure 3 Non-linear Canonical Correlation Analysis; sample distribution and attribute vectors

Legend					
		GRDCLEAR	Ground Clearance	SHLDLINE	Shoulder Line
		HEIGHT	Overall Height	SOFTTOP	Soft-Top (Canvas)
BMPRSTYL	Bumper Style	HLSHAP	Headlight Shape	TRIMSTYL	Side Trim Style
BNNTLEN	Bonnet Length	LEADIN	Lead-In Angle	WASHAPF	Fr. Wheel Arch Shape
BNNTSLOP	Bonnet Slope	LENGTH	Overall Length	WASHAPR	Rr. Wheel Arch Shape
BPILRAK	B-Pillar Rake	LHRATIO	Length/Height Ratio	WASIZEF	Fr. W-Arch Flare Size
BPILTKN	B-Pillar Thickness	PNLRNDN	Panel Roundness	WASIZER	Rr W-Arch Flare Size
CDPILRAK	C/D-Pillar Rake	ROOFLINE	Roofline Roundness	WASTYLF	Fr. W-Arch Flare Style
CDPILTHI	C/D-Pillar Thickness	RUNBOARD	Running Board	WASTYLR	Rr. W-Arch Flare Style
DOORS	Doors	SCRNRAKE	Screen Rake	WHEELBAS	Wheel Base Length
DROPOUT	Drop-Out Angle	SCRNRNDS	Screen Roundness	WIDTH	Overall Width
EDGERNDN	Edge Roundness	SHLDHGHT	Shoulder Height	WSTHGHT	Waist Line Height

A point of note is that the order of importance of the selected axes to differentiation, (as reflected in the regression coefficients), does not necessarily follow the amount of attribute variance explained by them in the principal components analysis. Some aspects which contribute greatly to perceptual differentiation may be encoded with a single variable (e.g., presence of a soft-top) and thus represent only relatively small amount variance in the attribute data. Conversely, perceptually less significant aspects may be represented by a range of covariant attributes, which would make them seem more important in any data-reducing multivariate analysis.

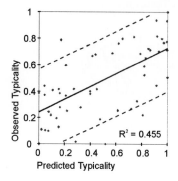

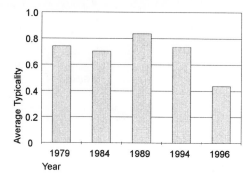

Figure 4 Observed vs. predicted typicalities

Figure 5 Average estimated typicalities by year

The pattern seen in the retained factors appears to be compatible with Prototype Theory, i.e., that some characteristics are more salient than others. From the regression coefficients, 'roundness', appears to be the most significant dimension, followed by 'proportion'. These two are fairly general factors, and it seems likely that 'roundness' and 'proportion' could be employed in differentiating between most motor vehicle groups.

The remaining two factors are concerned with more specific aspects; presence of a soft-top, and wheel arch emphasis. That a canvas top is perceived as significant is perhaps not surprising, given its major effect on a vehicle's profile, but the last included factor, wheel arch emphasis, is particularly interesting. Bearing in mind that no attributes concerning the wheels themselves were included in the analysis, this factor could be indicative of a general visual concern with the wheel area. Clearly, off-road capability is of interest in the RV product group, and it is not improbable that as a result some association is formed with the aesthetics of this area. Good off-road traction requires wider tyres, stability requires a broad track width, and long suspension travel results in large tyre to body clearances. All of these affect the styling in this area. That these latter two factors carry smaller weightings than the first two suggests that it is not until differentiation by the more general factors has been considered that attention switches to more specific aspects of styling.

The Homogeneity Analysis output shows a well-defined configuration. Sample points are distributed along three almost orthogonal straight lines, and there is a concentration of samples at what may be termed the 'nodes' of the distribution. These 'nodes' appear to represent some kind of 'cognitive focal centre' to each of the four clusters, in that many members of each cluster are positioned close to the node point. The tight grouping of these cluster members implies an especially strong common concept shared by the samples.

The presence of these nodes suggests that there exists a strong underlying structure of four conceptual groups perceived consistently by all subjects. Moreover, we can see that the extension of each concept contains certain samples that appear to be cognitively central to it, and that these samples are virtually always grouped together by all observers. As a corollary to this, it is apparent that these central members are never grouped with strong members of any other category, thus producing the concentration of samples seen around the 'node points'.

Distributed between these node points are the samples which appear less central to the conceptual groups, and which are prone to being grouped less predictably. This is not to say that a sample positioned close to, but not at, a node point is ever directly grouped with another concept. However, such a sample may be sometimes linked with other samples that are in turn associated with another category, in some kind of cognitive 'chain of association'. This chaining suggests that there are perceived associations which cross boundaries of clusters at a level below that of a superset of both clusters, i.e., that the perceptual groupings are not strictly hierarchical. These 'chains of association' then, result in the distributions found.

Figure 3 shows a similar sample distribution to that which was seen in the Homogeneity Analysis, although the orientation of the output space differs. The component loading vector plots for the attributes, however, show that there are very few single attributes that provide good discrimination. Nonetheless, some patterns do emerge.

In the plot of the first two axes we can see that only the lead-in angle (effectively a measure of the front overhang) approaches being a 'perfect' predictor. This probably results from the consistently large lead-in angle (i.e., short overhang) of the generic jeep types and lightweight RV type vehicles, which contrast with the consistently long overhang of the wagon type RVs. Other fairly good predictors are: the presence of a soft-top, which appears to be strongly indicative of jeep types, rear wheel arch size which contrasts RVs with both jeep and wagon types; and a collection of length-related attributes which contrast the heavyweight RVs and wagon types with the lightweight RVs and jeeps.

Inspection of the third axis, which mainly contrasts lightweight RVs with heavyweight RVs and jeeps, reveals even fewer strong predictors. To some extent a broad B-pillar seems to be characteristic of lightweight RVs., but perhaps most interesting is the bonnet length attribute. A proportionally long bonnet length appears to be an attribute shared by jeeps and heavyweight RVs, but not lightweight RVs. Whilst obviously related to engine size and configuration, it suggests that whilst both heavyweight and lightweight RVs maintain a visual association with the original jeep group, the attributes by which this association is made are different for the two RV groups.

In overview then, the most significant aspect is that, despite clear, generally agreed-upon perceived groups, there are no equally clear discriminating attributes that distinguish them. There is, in short, no 'necessary and sufficient' set of attributes to define any of these groups. Jeeps, for example, are strongly categorised by angularity, long bonnets and canvas soft-tops, and yet each of these properties can be found in members of at least one other group.

For example, many of the lightweight RVs also have soft-tops, and as already stated, a proportionally long bonnet is also a common feature of heavyweight RVs. Examples of angularity can be found in every other group (though more common in some than in others). Moreover, not all jeeps possess all three characteristics; some samples, for example, share only angularity out of these three variables. However, those vehicles at the 'focal centre' of the cluster generally share more of the common attributes associated with the group.

Similarly, this pattern emerges in other groups. Heavyweight RVs would appear to be differentiated from lightweight RVs mainly by length and proportion, and to a lesser extent by roundness, (heavyweight RVs tending to be more angular). However, not all members of the heavyweight group are long, and not all are angular. Inspection of individual cluster members showed, for example, that a number of only moderately long, two-door models have clustered firmly with the heavyweight group (which is predominantly made up of four-door models), but these shorter heavyweight models tend to be quite angular. Conversely, the more rounded members of the group are long, suggesting that there can be some kind of trade-off between attributes in meeting the criteria of cluster membership. Central members of the group, however, are both long and angular.

The plot of directly observed typicality judgements against the predicted values (figure 4), shows a moderate degree of correlation, although there is a good deal of scatter. Nonetheless, it appears that the attribute dimensions identified provide for some level of prediction of perceived typicality. Inspection of the outliers may indicate how the model may be refined by the inclusion of additional attributes to improve prediction.

Finally, it can be seen from figure 5 that the typicality of recent new vehicles tends to be significantly lower than that for previous years. Whilst these data only represent estimates of typicality as perceived at the time of collecting the data, if it is assumed that this is a representative 'snapshot' at any point in time, it suggests that new characteristics are perceived as unusual at first, but gradually come to be perceived as more normal over time as they are assimilated into a concept. Intuitively this does not seem surprising, and it is compatible with the view that successful new models generally emerge around the borders of existing product groups.

In conclusion, whilst a quantitative model for new product positioning would require a good deal more refinement of the techniques used here, it appears that there is good evidence for the useful application of a qualitative prototype theory model to the product design domain.

REFERENCES

Feger, H. and de Boeck, P., 1993, Categories and Concepts; Introduction to Data Analysis *In Van Mechelen et al.,* (eds.), Categories and Concepts: Theoretical Views and Inductive Data Analysis, (SanDiego, Academic Press).

Foxall, G. R., 1994, *Consumer Psychology for Marketing,* (Routledge, London) pp. 196-7.

Gifi, A., 1990, *Non-linear Multivariate Analysis,* (Chichester: John Wiley & Sons).

Hampton, J., 1993, Prototype Models of Concept Representation In *Van Mechelen et al.,* (eds.), 1993, Categories and Concepts: Theoretical Views and Inductive Data Analysis, (San Diego: Academic Press).

Lakoff, G., 1987, *Women, Fire, and Dangerous Things,* (Chicago: University of Chicago Press).

Rosch, E. and Mervis, C. B., 1975, Family resemblances: studies in the internal structure of categories. *Cognitive Psychology,* **7**: pp. 573-605.

Rosch, E., 1975, Cognitive representations of semantic categories. In *Journal of Experimental Psychology: General,* **104**: pp. 192-232.

Shackleton, J. P. and Sugiyama, K., 1996, Analysis of Trends in Japanese Recreational Vehicle Design. *Bulletin of the Japanese Society for the Science of Design,* 42, 6, 19-28.

Urban, G. L., Hulland, J. S. and Weinberg, B. D., 1983, Premarket Forecasting of New Consumer Durables: Modelling, Categorisation, Elimination, and Consideration Phenomena. In *Journal of Marketing,* **57**.

Directing Designers towards Innovative Solutions

Heleen Snoek and Paul Hekkert

This paper demonstrates that restructuring a design problem into user-product relations by involving knowledge from a variety of domains – even domains not directly related to the design problem at hand, determines to a large extent the path to innovations. Designing is considered a real-world and productive (creative, non-routine) problem solving activity. As in any creative process, problem definition is viewed as the most important component, because it defines the solution space in which the problem will be solved. A solution space based on familiar knowledge structures might restrain the designer from arriving at original design solutions. In order to come up with an innovative design, the designer must redefine the initial problem into a novel one by restructuring existing knowledge and breaking through constraining boundaries.

In an experiment, student designers were instructed to go beyond their initial solution space during the design of an alarm clock for the year 2002. Their concepts were compared to concepts designed by a control group that received no such instructions. The concepts were judged by design experts and rated for originality and appropriateness. After controlling for problem solving style, specification level of concept and quality of presentation, significant differences were found between both groups. The alarm clocks designed by the experimental group were more creative (original and appropriate) than those designed by the control group.

This result implies that directing designers to the (future) consumer product relationship given a set of (future) conditions, can result in innovative design solutions.

INTRODUCTION

Innovation and design

From consumer research (e.g., Garber, 1995; Schoormans and Robben, 1997) we know that a product showing off for its visually typical, original or novel appearance draws attention and that such a product will be evaluated positively when it is judged as appropriate for the goal. Moreover, innovation occurs when a breakthrough is reached in the development of a technology or product-line. Independent of the label attached to it, deviation from familiar stimuli requires creativity. Hence, an interesting question for inventors and designers is how to produce creative products.

Successful innovations often seem to come into existence out of the blue, however, reality tells us that "Human invention never produces something entirely out of nothing. .... [Therefore,] Invention is better seen as an act of combination rather than an act of *ex nihilo* creation." (Perkins, 1988). In their attempts to produce creative solutions, designers make plans, drawings or models of something in order to decide how it will look, work etc. Their decisions are based on what they know from experience or what they can find in a certain product related domain. The underlying amount of knowledge and the way this information is structured, determines to a large extent whether a new design will be innovative or just a new product in a larger line of similar products. In other words, the right combination and treatment of knowledge can produce a breakthrough.

Of course, it can not be known in advance what 'the right combination' is. However, in this paper a strategy is presented that raises the probability of finding a right combination and in this way can lead to creative designs.

Design, problem solving and solution spaces

In order to understand the relationship between knowledge and design, we, among others (e.g., Newell and Simon, 1972; Thomas and Carroll, 1979; Lawson, 1980), consider designing a real-world and creative, 'non-routine' problem solving activity. Non-routine problems are problems for which the problem solver must invent a new way to solve (Mayer, 1995). In order to produce a new solution, problem solvers have to restructure the problem, i.e. they must become aware of new relations among problem components. Such activities are called 'productive thinking' (e.g., Dominowski, 1995), which is closely related to creativity.

Analogous to non-routine problem solvers, designers face new situations, e.g. new consumer needs, for which they have to find a new, creative (design) solution.

In order to solve a specific problem a certain body of knowledge is needed. The total amount of knowledge related to a certain problem and including both the initial state of knowledge and the goal state of the problem, as well as a set of operators that can produce new states of knowledge from existing states of knowledge, is metaphorically called the 'problem space' (Newell and Simon, 1972) or the 'solution space' (Goldschmidt *et al.,* 1996). 'Solution space' seems to be a more appropriate concept for it covers all possible solutions that the problem solver might consider. Boundaries of a solution space are set by constraints due to (limited) experience, knowledge or preconceptions of the problem solver.

The solution space of a design problem involves knowledge from a variety of domains. The body of knowledge and its interrelations present in and directly accessible to the designer, e.g., retrieved from long term memory while studying the design brief, can be described as the initial solution space (de Vries, 1990). However, the complexity of most design problems requires gathering additional information, either from domains closely related to the problem at hand or even from domains that have never been related to this particular problem before. Thus, the final solution space expands from the initial solution space and includes information extant in memory and knowledge learned from new information (see Figure 1). Productive thinking, i.e. restructuring the information in the finalsolution space, will lead to creative solutions to the novel (design) problem (Mumford *et al.,* 1994).

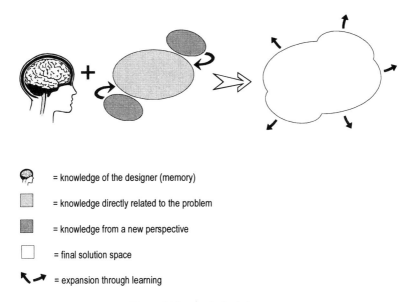

= knowledge of the designer (memory)

= knowledge directly related to the problem

= knowledge from a new perspective

= final solution space

= expansion through learning

Figure 1 Metaphorical solution space

From theory to practice

At our faculty students learn a rational-analytic design process, which implies roughly that divergent thinking precedes convergent reasoning (Roozenburg and Eekels, 1995). Practically it means that they learn to gather information concerning user-context, comparable products and style characteristics in order to build up their solution space. Then possibly relevant items are selected (constraints are set) in order to formulate requirements.

Unfortunately, designers are not always used to looking for information in other disciplines than those closely related to the design problem at hand (Christiaans, 1993). Undeliberately they constrain their solution space and thus will limit their output of conceptual design. This phenomenon is called functional fixedness or fixation.

For designers, it is important to be aware of the possibility to become fixated and to be able to overcome such a state, for fixation results often only in incremental developments only. Recent research indicates that a product centred formulation of the design problem (Goldschmidt *et al.*, 1996) might constrain the solution space as well. Restructuring the design problem in terms of user-product interactions (e.g., 'design a device that will ease a paver's knee ache' instead of 'design a knee protector for a paver') might help to overcome the fixation problem, because the problem is seen in a broader view.

Also, fixation effects can occur when examples of familiar design solutions are taken into account. Designers, searching for new solutions in familiar solution spaces, will come up with solutions similar to the example (Gielen, *et al.*, 1998; Gero, 1996). Searching for solutions in remote domains, however, enables viewing the problem from a new perspective, which might result in original design solutions or even a 'quantum leap' forward in a certain domain.

Original solutions are not always appropriate solutions. Generating a creative (original and appropriate) or an innovative solution requires the awareness of the designer that products are made to fulfil human needs and that those needs and possible product characteristics change over time due to changing conditions affected by, e.g., technological, social, political or cultural aspects (Hekkert, 1997). From these theoretical considerations it is hypothesised that directing designers to take a (novel) context as a point of departure for their design process will increase their ability to fundamentally restructure a familiar design problem and to generate innovative design solutions.

In order to test this hypothesis, an experiment was carried out in which a number of the subjects were challenged to go beyond the initial solution space.

METHOD

Subjects

Subjects were 30 students (mean age 22.3 years, range 21 to 27 years) from the Subfaculty of Industrial Design Engineering of Delft University of Technology. Students were allowed to participate in the experiment when they completed the course 'Design-assignment IV' (of 6 regular design-assignments), and were not yet involved in their final graduation project. Each subject was paid an equivalent of 8 hours work, irrespective of the total time they spent on the design task.

Design task and procedure

The subjects had to design two concepts of an alarm clock for the year 2002 (at that time this was 5 years ahead). The subjects were asked to work individually. The experiment was divided into a preparation stage and an experimental session.

The preparation stage started with an instruction session (about half an hour, individually) in which the experimenter explained the design problem. After this oral instruction subjects took a written version of the instruction home. From then they were supposed to prepare the design problem in their own time. Literally they were asked to gather sufficient information to be able to generate two concepts for an alarm clock for the year 2002.

During this preparation stage each student had to write down the activities and the time they had been spending on them. In the instruction it was stressed that 'thinking' was a design activity as well, in order to make sure the subjects reported their activities as complete as possible. The mean preparation time turned out to be 322 min (SD: 184 min; range: 105 to 885 min).

Three days after the instruction session the subjects came back for the experimental session (about 3.5 hours, individually) where they had to generate two concepts for the alarm clock. After arrival in the experimental room, the subjects first were interviewed about their progression. The questionnaire covered a number of items concerning their search for information. The interview was taped. After this questionnaire they could start sketching and working out the two concepts. It was stressed that the basic idea of the alarm clock should be clear to outsiders, therefore the subjects were asked to add comments to the drawings.

Every 30 minutes the subjects had to fill out a small form concerning their activities (reading, thinking, sketching etc.) and the time they spent on it. After 1.5 hours the experimenter warned the subject that half of the design time had passed. After 2.5 hours the experimenter asked the subjects to bring the concepts to an end. The subjects were required to finish their work within 3 hours, but no minimum time was set. Only one subject did not manage to produce a second concept. As a result 59 concepts for an alarm clock for the year 2002 were produced.

Independent variables

Experimental manipulation

Although all subjects had to design an alarm clock for the year 2002, the subjects were randomly assigned to one of two conditions, a control condition and an experimental condition. The subjects in the control-group (7 females, 7 males) were not explicitly instructed to apply a particular design method. The subjects in the experimental group (8 females, 8 males) were given additional information concerning the relationships between human needs, the product and the context in which the consumer and the product will interact. Further, it was stressed that human needs change over time due to changing contexts. The subjects were asked to keep these ideas in mind while designing, and to form a mental picture of the environment in which their alarm clock would interact with its consumers.

Cognitive style

The way in which problems are solved differs between people and depends on the cognitive style a person is likely to use. There is a variety of theories concerning cognitive styles (Sternberg and Grigorenko, 1997), one of them is Kirton's adaption - innovation theory (Kirton, 1976). Not too much is known about the effect of cognitive style on the originality of design solutions. From creativity studies it is learned that adaptors and innovators both can be original persons, they merely differ in their approach to (design) problems rather than in their final achievements (Goldschmidt, 1994).

Because this experiment studies the effect of different approaches to design problems, non-equal distribution of cognitive styles in the two conditions might affect the results undesirably. It was therefore decided to have the subjects fill out Kirton's standard questionnaire in the instruction session in order to obtain their KAI-scores. An ANOVA (analysis of variance) was used to test whether the male and female adaptors and innovators were equally distributed over the two groups.

We found a normal distribution of KAI-scores in our group of subjects reflecting the distribution found by Kirton (1976) in a population of 532 subjects. Mean KAI-score for our group of subjects was 95.6 (theoretical mean is 96), $SD = 18.2$; and the scores ranged from 59 to 128 (theoretical range = 32-160). In accordance with Kirton (1994), females scored significantly lower than males (87.4 vs. 104.3, $F_{(1,26)} = 8.59$, $p < .01$). Only a slight difference was found for condition ($F_{(1,26)} = 3.24$, $p = 0.08$): KAI-scores of subjects in the control group were higher ($M = 101.4$, $SD = 19.9$) than in the experimental group ($M = 91.0$, $SD = 15.6$).

These findings support our decision that KAI-scores should be controlled for.

Specification level of concepts

The subjects were asked to design two concepts for an alarm clock, where concept was defined as "a materialised idea, in which the first signs of design are observable, but details are still lacking. It should be able to derive the design features for the final product". A quick glance at the 59 sketches showed that there still was a large variety in the specification level of these 'concepts'.

Sketches that are specified to different levels are hard to compare objectively for underlying qualities. Therefore, an indication of the specification level for each concept was needed in order to control for these undesired differences. Based on expert judgements the specification level was obtained for each concept in order to control for this variable.

Quality of presentation

Another undesired difference between the 59 concepts concerned drawing skill. The quality of the concept is supposed to be independent of the quality of presentation, but the latter easily affects the perception in judging. Non-equal distribution of the quality of presentation in the control group compared to the experimental group might affect the results undesirably. Therefore, it was decided to statistically control for quality of presentation, a measure that was also obtained by expert judgements.

Dependent variables

A total of twelve qualitative characteristics of each of the concepts was assessed on a seven-point bipolar scale by ten judges rating independently. The judges were experienced designers of whom seven work in educational environments (from our own as well as from other design institutes) and three work at design studios. Two series of 30 concepts (a fake concept replaced the missing one), including either the first or the second concept of each designer, were presented randomly to each of the judges separately. They spent on average three and a half hours on the task.

For each concept the procedure was as follows. A series of 30 concepts was spread on tables in a room. Each judge was allowed to have a look at the sketches in order to get an impression of the general level. Then the judge had to start with rating the first concept.

The very first scale was a general question about the specification level of the concept (control variable). The next eight scales covered scales measuring originality (not original - original, typical - atypical), attractiveness (not likely to be trendsetting - likely to be trendsetting, unattractive - attractive), appropriateness for the year 2002 (not fulfilling needs - fulfilling needs, inappropriate solution - appropriate solution) and coherence (not coherent - coherent (form), not transparent - transparent (idea)). Then a short description of the subject's ideas and considerations about the year 2002 as derived from his/her notes and from the interviews was presented to the judges. With this in mind the judge had to rate two more scales concerning appropriateness (reflecting the ideas as formulated by the designer - not reflecting the ideas as formulated by the designer, inappropriate solution - appropriate solution (when knowing the ideas of the designer)). The final scale was a general question about the quality of presentation (control variable).

Scores on coherence will not be considered in this paper.

RESULTS

Manipulation check

To test the effectiveness of the manipulation and to ensure that subjects in the experimental condition did in fact look for information in other disciplines other than those closely related to the design problem at hand, the reported activities in the preparation stage were compared for both groups. ANOVAs were used to analyse the effects of the manipulation on the invested time and gathered information.

In line with the method our students have learned, it was expected that the subjects in the control group mainly gathered information about current waking devices, waking habits and sleeping rooms. The extra information given to the subjects in the experimental group was intended to direct them to search for information from other domains - especially contextual domains - as well. Since such information is less easily available, it is also predicted that these subjects would invest more time gathering information.

Invested time

From the logbook we could derive the time they spent on the activities at home. The main activities concerned looking for information (29 subjects), thinking (29 subjects), analysing (17 subjects), sketching (17 subjects), making a collage (9 subjects), and discussing with others (6 subjects). A significant effect was found for the time they spent on the preparation stage, $(F_{(1,28)} = 4.55, p < .05)$. Subjects in the experimental group took longer preparation time ($M = 385$ min, $SD = 218$ min) than subjects in the control group ($M = 249$ min, $SD = 102$ min). It is interesting to notice that this difference is mainly caused by the time spent on gathering information $(F_{(1,28)} = 4.28, p = .05)$. Subjects in the experimental group searched significantly longer for information ($M = 167$ min, $SD = 122$ min) than subjects in the control group (M = 92 min, SD = 68 min). The time spent on other activities did not differ significantly between the groups.

These results confirm our prediction that subjects in the experimental group invested more time gathering information.

Information gathered

An inventory of the information the subjects reported on was made. As expected many subjects came up with information about current waking and other (electronic) devices. Also brochures concerning current contexts (e.g. lifestyle magazines) were popular. Only four of the subjects in the control group reported the use of information beyond waking related topics, whereas twelve subjects in the experimental group came up with information from a broader perspective. Chi-square analysis showed that this difference is significant (χ^2 = 6.47, p < .05).

This result indicates that the manipulation had the desired effect: subjects in the experimental condition more often built up a solution space covering a larger variety of knowledge sources compared to subjects in the control condition.

Interjudge reliability and concept independence

Interjudge reliability

Ten judges were involved in rating the concepts. In order to check whether these judges agreed with each other, a reliability analysis was carried out. The reliability coefficient (α-value) varied between 0.66 (for Specification Level of Concept) and 0.93 (for A-typicality). Further, the reliability coefficient was analysed in order to find out whether one (or more) particular judge might have suppressed it. It turned out that one of the judges disagreed with the other judges on eight of the twelve scales (Table 1). Therefore we decided to eliminate that judge and based the final scores of the concepts on the ratings of the 9 remaining judges.

Concept independence

By asking two concepts of each subject we had to decide how to handle these data. If scores on concept 1 were highly correlated to scores on concept 2 we could compute the mean score of the two concepts for each subject. However, the correlation between concept 1 and concept 2 was only significant for Quality of Presentation (r = 0.59, p < .001). For the remaining eleven scales, correlation's ranged from 0.06 (for Fulfilling needs) to 0.33 (for A-typicality). So, ratings of concept 1 turn out to be highly independent of ratings of concept 2. Analyses will therefore be carried out over the total set of 59 concepts.

Table 1

Relative reliability if one judge is eliminated

scales->	SoC	Ori	Coh	Ful	Att	Aty	Tra	ApA	Tre	Ref	ApB	QoP
judge 1	-	-	-	-	-	-	-	-	-	-	-	-
" 2	-	-	+	-	-	-	-	-	-	-	-	+
" 3	-	-	-	-	-	-	-	-	-	-	-	-
" 4	-	-	-	-	-	-	-	-	-	-	-	-
" 5	-	-	-	-	-	-	-	-	-	-	-	-
" 6	+	+	-	-	-	+	-	-	-	-	-	+
" 7	-	-	-	-	-	-	+	-	-	-	-	-
" 8	-	-	+	+	+	-	+	-	+	+	+	+
" 9	-	-	-	-	-	-	-	-	-	-	-	-
" 10	-	-	-	-	-	-	-	-	-	-	-	-
a (10)	.66	.91	.74	.73	.74	.93	.78	.71	.70	.77	.65	.85
a (9)	.63	.91	.74	.73	.77	.93	.79	.70	.71	.78	.71	.85

Note: SoC = specification level of concept; Ori = originality; Coh = coherence of form; Ful = fulfilling needs; Att = attractivity; Aty = atypicality; Tra = transparency of idea; ApA = appropriateness; Tre = degree of trendsetting; Ref = reflecting ideas of the designer; ApB = appropriateness when knowing the ideas of the designer; QoP = quality of presentation; + = reliability coefficient increases if judge is eleminated; - = reliability coefficient declines if judge is eliminated.

Results

The final dataset consisted of mean scores over 9 judges on 10 scales for 59 concepts, including 2 control scales to be used as covariates in the primary analyses.

Each characteristic of interest, i.e. originality, attractiveness, appropriateness before and appropriateness after knowing the ideas of the subject, consisted of two scales. It would be appropriate to examine correlations between each of the scales before conducting the primary analyses. Table 2 presents the means, standard deviations, and correlations for all variables. As predicted, Originality and A-typicality tend to be strongly related to each other ($r = 0.86$). Attractiveness and Trendsetting however, turned out to be less strongly related ($r = 0.41$).

Attractiveness correlates well with Originality, Fulfilling Needs and Appropriateness ($0.73 < rs < 0.83$) and thus can be seen as a general measure of creativity. As predicted, Fulfilling Needs is strongly correlated to Appropriateness ($r = 0.84$), and Reflecting the ideas of the designer and Appropriateness when knowing the ideas of the designer are also strongly correlated ($r = 0.83$).

Table 2

Means, Standard Deviations, and Correlations for Dependent Variables and Covariates

	M	SD	KAI	SoC	Ori	Ful	Att	Aty	ApA	Tre	Ref	ApB	QoP
KAI	95.6	18.2	-	.11-	.15	.09	.13	.08	.01	.05-	.07-	.07	.00-
SoC	2.92	0.5		-	.14	.38**	.36**	.05-	.45**	.48**	.25	.33*	.75**
Ori	3.76	1.3			-	.72**	.82**	.86**	.52**	.04	.29*	.53**	.30*
Ful	3.52	0.8				-	.80**	.51**	.84**	.50**	.47**	.68**	.42**
Att	3.26	0.9					-	.61**	.73**	.41**	.38**	.71**	.43**
Aty	3.85	1.4						-	.33*	.13-	.21	.40**	.18
ApA	3.56	0.8							-	.76**	.53**	.75**	.45**
Tre	3.46	0.9								-	.50**	.60**	.27*
Ref	3.73	0.9									-	.83**	.24
ApB	3.60	0.8										-	.37**
QoP	3.12	0.8											-

Note: KAI = KAI-score; SoC = specification level of concept; Ori = originality; Ful = fulfilling needs;

Att = attractivity; Aty = atypicality; ApA = appropriateness; Tre = degree of trendsetting;

Ref = reflecting ideas of the designer; ApB = appropriateness when knowing the ideas of

the designer; QoP = quality of presentation

*p < .05, **p < .01

The primary analyses consisted of a series of ANOVAs used to analyse the effect of experimental manipulation on the originality, attractiveness and appropriateness of the design solutions. In these primary analyses KAI-scores, Specification Level of Concept and the Quality of Presentation were included as covariates (see Table 3).

Originality

A significant effect was found for Originality ($F_{(1,54)} = 4.52, p < .05$). The concepts of the experimental group scored higher on originality ($M = 3.87, SD = 1.2$) than the concepts of the control group ($M = 3.64, SD = 1.4$). Also a significant effect was found for A-typicality ($F_{(1,54)} = 4.65, p < .05$). The concepts of the experimental group were more a-typical ($M = 4.03, SD = 1.3$) than the concepts of the control group ($M = 3.63, SD = 1.5$).

Attractiveness

The concepts designed by the experimental group ($M = 3.29, SD = 0.9$) turned out to be slightly but not significantly more attractive ($F_{(1,54)} = 3.35, p = .07$) than those designed by the control group ($M = 3.22, SD = 0.9$). There were no significant effects found for Trendsetting ($F_{(1,54)} = 0.44$).

Table 3

Analyses of variance summary table for Experimental condition

Source	concept 1 and 2 (n=59)		concept 1 (n=30)	concept 2 (n=29)
	(a) KAI/SoC/QoP	(b) Original	(c) KAI/SoC/QoP	(d) KAI/SoC/QoP
	$F (1,54)$	$F (1,57)$	$F (1,25)$	$F (1,24)$
1) Originality	**4.52***	0.46	1.86	**3.18**
2) A-typicality	**4.65***	1.12	1.31	**4.26***
3) Attractiveness	**3.35**	0.10	2.83	0.93
4) Trendsetting	0.44	0.14	0.57	0.18
5) Adapted	1.53	0.00	1.33	0.93
6) Appropriateness	0.34	0.00	0.51	0.05
7) Reflection of designer's ideas	0.60	0.17	**3.27**	0.19
8) Appropriateness (when knowing..)	**3.21**	0.92	**4.34***	0.25
9) Creativity (1,2,5,6)	**4.28***	0.40	1.75	2.50
10) Creativity (1,2,7,8)	**5.05***	0.98	**3.74**	2.27

Note: (a) = primary ANOVA with covariates KAI-score, Specification level of Concept and Quality of Presentation

(b) = original ANOVA (c) = as (a) concept 1 separately (d) = as (a) concept 2 separately

(F-values with p <0.1 are printed in bold face, significancies are flagged)

* p < .05

Appropriateness

Two kinds of appropriateness were distinguished: appropriateness before knowing the ideas of the designer and appropriateness when knowing the ideas of the designer. The scales that were thought to measure appropriateness were Adaptation to waking-needs, Appropriate solution in 2002, Appropriate solution when knowing the ideas of the designer, and Reflecting the ideas of the designer.

A slight tendency was found for Appropriateness when knowing the ideas of the designer ($F_{(1,54)} = 3.21$, $p = .08$). Subjects in the experimental group designed more appropriate concepts ($M = 3.68$, $SD = 0.8$), than those in the control group ($M = 3.49$, $SD = 0.7$).

Creativity

Creativity can be defined as both original and appropriate (Mumford, *et al.,* 1994). Therefore, we cumulated the originality and appropriateness scores. For we distinguished between appropriateness before and after knowing the ideas of the designer, two cumulative creativity scales were made. The results of the ANOVAs showed a significant effect for the creativity scale before the judges knew the ideas of the designer ($F_{(1,54)} = 4.28$, $p < .05$). Concepts designed by the experimental group were rated more creative ($M = 14.97$, $SD = 3.5$) than those designed by the control group ($M = 14.35$, $SD = 4.0$). The effect was even stronger when the judges knew the ideas of the designer ($F_{(1,54)} = 5.05$, $p < .05$): the design solutions of the experimental group were rated more creative ($M = 15.36$, $SD = 3.6$) than those in the control group ($M = 14.43$, $SD = 3.6$).

DISCUSSION AND CONCLUSION

The present results corroborate the main hypothesis of this study. Student's who are simply instructed to map a future context for a human-product interaction and take this context as a starting point for their design, come up with more creative solutions than students who received no such instruction. The analysis of the student's activities in the preparation stage revealed that such instruction stimulated them to spend more time on searching for information from domains further away from the target domain. This information will have enabled them to enlarge their solution space and thereby resulted in the generation of more creative, i.e., original and appropriate, design solutions.

Although our hypothesis could generally be confirmed, this result might have been caused by the way we treated the data. As stated before of concept 1 turned out to be uncorrelated to ratings of concept 2 (except for quality of presentation) and we therefore decided to carry out our analyses over the total set of 59 concepts, i.e., as if all concepts were independently developed. In order to determine whether and how this treatment of the data affected the results, we also analysed the scores of both concepts separately.

Not surprisingly, the second concept received significantly lower mean scores on all scales than the first concept. It can generally be concluded that the students put most effort in the first concept. With respect to the differences between the two conditions, separate analyses of both concepts lead to major differences in the results (see Table 3, columns c and d). ANOVAs for each scale based on the scores of concept 1, revealed a significant effect for Appropriateness when knowing the ideas of the designer ($F_{(1,25)} = 4.34$, $p < .05$) and a weak, albeit not significant effect for Reflection of the designer's ideas ($F_{(1,25)} = 3.27$, $p = .09$). First concepts designed by the experimental group are rated as more appropriate and better reflect the designer's ideas than those by the control group. These two effects completely disappear when the analyses are based on the scores of the second concept. Instead, based on these concepts, subjects in the experimental group obtained higher ratings on A-typicality ($F_{(1,24)} = 4,26$, $p < .05$) and originality ($F_{(1,24)} = 3.18$, $p = .087$).

The combined results of these separate analyses therefore show that our instruction increased the appropriateness of the first concepts (given that the designer's ideas are known to the raters) but not their originality and lead to relatively more original second concepts which were however not more appropriate. The latter finding might tentatively be explained by the broader context, i.e., an enlarged solution space that offered the experimental group more 'space' to generate original solutions. Why this context did not result in more original first concepts is difficult to explain, but might have been caused by the subjects' fixation on making their concepts first of all a clear reflection of their ideas without being focused on novelty per se. This tendency is then reflected in the higher appropriateness scores.

Overall, it seems that the students in the experimental group were too inexperienced to translate their newly developed context into a design that is both original and appropriate. As a result, scores on creativity, in being based on the sum of the originality and appropriateness scales, did not significantly differ between both groups when analysing the concepts separately.

However, for both measures of creativity, the effects were significant when the analyses are performed over all 59 concepts. Still, these effects were not very strong. Next, table 3 clearly reveals the predicted effects could only be observed by including the three control variables as covariates into the ANOVAs (column a). Analysis of variance tests without covariates (column b) yielded no significant effects whatsoever. We will conclude with discussing some tentative explanations for the relative weakness of the effects obtained.

First, the assignment 'design an alarm clock for the year 2002' might have had a negative effect on the magnitude of the effects. Since all subjects were tuned to design a product for the (near) future, even the subjects in the control group might have been tempted to take future developments into account. On the other hand, 2002 may be too short a term to expect major changes in sleeping habits and waking devices. Except for the transition from mechanical alarm clocks to electronic clock radios, this product has not undergone major changes in the last few decades.

A second explanation for the weakness of the findings stems from personal observations by the authors. Examination of the information reported and ideas formulated by the subjects in the experimental group indicated that most of them indeed managed to create a novel context from where they could define new user-product interactions. Only a very few of them were however able to translate that context into their design solution. Even if they have the required material, it seems to be difficult for these students to define new interactions between the consumers and the product, which might lead them to original design solutions. Again, this observation suggests that the students generally lacked the experience to work successfully with this design approach. To apply this approach in future design projects, more training seems to be recommended (see Hekkert, 1997).

Finally, it turned out that the final concepts differed considerably as to their specification level and the quality of the presentation. Although we have statistically controlled for these two variables, we believe that the judgements of the experts would have been more reliable when all concepts were sketched by one drawer (see Dorst, 1997).

Despite the procedural limitations mentioned, it is clear that it is possible to direct designers to creative or innovative solutions. In future research the procedure used in this experiment must be refined in order to find the most relevant aspects of the proposed directing strategy.

REFERENCES

Christiaans, H.H.C.M., 1993, The Effects of Examples on the Use of Knowledge in a Student Design Activity: The case of the 'Flying Dutchman'. In *Design Studies,* **14**, pp. 58-74.

Dominowski, R.L., 1995, Productive Problem Solving. In *The Creative Cognition Approach,* Smith, S.M., Ward, T.B. and Finke, P.A., (Eds.), (Cambridge, Mass: MIT Press), pp. 73-95.

Dorst, K., 1997, *Describing Design: A Comparison of Paradigms.* Unpublished doctoral thesis. Delft University of Technology.

Garber, L.L., 1995, The Package Appearance in Choice. In *Advances in Consumer Research,* **22**, pp. 653-660.

Garber, L.L., 1995, The Package Appearance in Choice. In *Advances in Consumer Research,* **22**, pp. 653-660.

Gielen, M.A. Hekkert, P. and van Ooy, C.M., 1998, Problem Restructuring as a Key to a New Solution Space: An Example Project in the Field of Toy Design for Disabled Children. In *The Design Journal,* in press.

Goldschmidt, G., Ben-Zeev, A. & Levi, S. 1996, Design Problem Solving: the effect of problem formulation on the construction of solution spaces. In *Cybernetics and Systems* '96: Vol. 1, Trappl., R. (Ed.), (Vienna: Austrian Society for Cybernetic Studies), pp. 388-393.

Goldschmidt, R.E., 1994, Creative style and personality theory. In *Adapters and Innovators: Styles of Creativity and Problem Solving.* Kirton, M.J. (Ed.), revised edition, (London: Routledge), pp.34-50.

Hekkert, P., 1997, *Productive designing: A path to creative design solutions.* Proceedings of the Second European Academy of Design Conference. Available Internet: http://www.svid.se/ead.htm.

Kirton, M.J., 1976, Adaptors and Innovators: A Description and a Method. In *Journal of Applied Psychology,* **61**, pp. 622-629.

Kirton, M. J., 1994, *Adaptors and Innovators at Work.* In Kirton, M. J., (Ed.), Adapters and Innovators: Styles of Creativity and Problem Solving, Revised edition, (London: Routledge), pp. 51-71.

Lawson, B., 1980, *How Designers Think.* (London: The Architectural Press).

Mayer, R.E., 1995, The Search for Insight: Grappling with Gestalt Psychology's Unanswered Questions. In *The Nature of Insight.* Sternberg, J. and Davidson J. (Eds.), (Cambridge: MIT Press), pp. 3-32.

Mumford, M.D., Reiter-Palmon, R. and Redmond, M.R. 1994, Problem Construction and Cognition: Applying Problem Representations in Ill-defined Domains. In *Problem Finding, Problem Solving, and Creativity.* Edited by Runco, M.A., (Norwood, NJ: Ablex), pp. 3-39.

Newell, A. and Simon, H.A. 1972, *Human Problem Solving.* (Englewood, Cliffs, NJ: Prentice Hall).

Perkins, D.N., 1988, The Possibility of Invention. In *The Nature of Creativity: Contemporary Psychological Perspectives,* Edited by Sternberg R.J., (Ed.), (Cambridge: Cambridge University Press), pp. 362-385

Purcell, A. T. and Gero, J. S. 1996, Design and Other Types of Fixation. In *Design Studies,* **17**, pp. 363-383.

Roozenburg, N. F. M. and Eekels, J. 1995, *Product Design: Fundamentals and Methods.* (Chichester: John Wiley & Sons).

Schoormans, J. P. L. and Robben, H. S. J. 1997, The Effect of New Package Design on Product Attention, Categorization and Evaluation, In *Journal of Economic Psychology,* **18**, pp. 271-287.

Sternberg, J. and Grigorenko, E.L. 1997, Are Cognitive Styles Still in Style? In *American Psychologist,* **52**, pp. 700-712.

Thomas, J. C. & Carroll, J. M. 1979, The psychological Study of Design. In *Design Studies,* **1**, pp. 5-11.

Vries, E. de, 1990, *Problem Solving in Design: A Comparison of Three Theories,* OCTO-report 90/03, Eindhoven University of Technology.

Networking New Product Development: the integration of technical and product innovation

Anne Tomes, Peter Armstrong and Rosie Erol

Using case study material from an advanced chemical technology, this paper identifies two chains of connection between the developer / producer of a new technology and the manufacturers of consumer products on the other. In the first, the connection is direct, whilst in the second it is effected through an intermediary company which has developed an expertise in tailoring the new technology to particular applications. Whilst the direct connection has resulted in a number of New Product Development (NPD) failures, the mediated connection has been consistently successful. Through an examination of a representative NPD programme within each chain of connection, the paper examines the reasons for these differential success rates, and discusses the implications for the organization of NPD within advanced technology fields.

INTRODUCTION

This paper is concerned with the distinction between technological and product innovation, and with the effectiveness of different forms of inter-company network in achieving the latter. The idea that the two forms of innovation may best be achieved through different forms of co-operation is not entirely new. An exploratory study by Moenaert and Caeldries, (1996) suggests that technological learning may be improved by the facilitation of internal communication flows within a Research and Development (R&D) team whereas market learning may be advanced by contacts with other R&D teams. The present (equally exploratory) study compares a representative product development programme carried out within a company specialising in the development and production of an advanced technology with one carried out by a company specialising in applications of the same technology (called 'specialist chemicals' in this paper for reasons of confidentiality). Though the programme in the producer company achieved a measure of technological success, it was, nevertheless, a failure in the market-place. The one carried out in the applications company, in contrast, was a success, despite involving less in the way of technological development. The assumption behind the unsuccessful programme appears to have been that the solution of a series of technological problems would automatically add up to an innovative product, and its failure shows the importance of maintaining an end-user's, as opposed to a technologist's perspective on innovation.

A second, related, objective of the paper is to explore the effectiveness of different forms of inter-company network in the NPD process, particularly in its crucial early stages (Cooper, 1988). The advantages and problems of networks as a means of pooling resources of technology, intellectual property and finance have been extensively discussed in a variety of high-technology industries - for example, pharmaceuticals (Bower, 1993), computing (Noren *et al.,* 1998), chemical processing (Hutcheson *et al.,* 1995) and home automation (Tidd, 1995). This body of work, however, has yet to produce an accepted vocabulary for different patterns of inter-company interdependence. Consequently the question of which pattern might most effectively promote NPD has scarcely been posed. The present paper offers a beginning in this respect, since the data suggest that the relationship between a company specialising in technological development and the manufacturers of consumer products may be more effective when mediated by a company specialising in applications of the technology.

The two themes of the paper are connected in that the respect in which the mediated relationship was superior to the direct one was in the integration of product and technological development. In the mediated relationship, product concepts originated with consumer goods manufacturers rather than with either the developer of the technology or the application specialist, thus ensuring that due weight was assigned to the product innovation aspect. In this connection, survey data by Karakaya and Kobu, (1994) show the advantages of using customers as the source of new product ideas, whilst the importance of involving customers in product design at a more general level has been confirmed by Gemunden, *et al.,* (1992), Shaw, (1993) and Hutcheson, *et al.,* (1996).

CASE STUDIES

Multichem

Multichem is a large multinational company that produces a range of products within the chemical industry. The UK company has an annual turnover of around £100 million, and employs about 1000 people.

Currently, a Specialist Chemical Group within the Industrial Chemicals Division carries out the development and production of specialist chemicals. This group, consisting of about a dozen research staff plus the marketing and production teams, is responsible for the development of consumer applications as well as the technology itself. We talked with staff for a total of seven hours, with the data coming from taped interviews with the Business Development Manager of the Specialist Chemical Group, (quoted below).

It will be evident that Multichem's early presence in the market for specialist chemical applications was built on successful R&D programmes. Perhaps for this reason, and perhaps because the supply of scientific apparatus to universities and colleges is still a major part of the company's business, there remains a strong element of R&D - push in Multichem's search for consumer markets:

"... Therefore a lot of new products come out of R&D. It's people playing with the vivid effects, and thinking we must be able to do something with this. And then it's given to the marketing department to do something with it."

Unlike research laboratories, however, commercial customers could not be expected to master the delicacies of handling the specialist chemicals, since they are very sensitive to contamination. If the vivid effects were to be exploited in commercial applications, it was necessary to find some means of insulating the specialist chemical component from other components of the final mixture. The answer developed over four years' R&D work was to process the specialist chemicals further. This solution therefore, represented a major R&D investment for Multichem before the search for consumer applications could even begin: "We were making specialist chemicals, and then because everyone said they found them so difficult to use, we took it on to the next stage ourselves, and started encasing them. Many other people who have wanted to use these wonderful effects couldn't do the modification. Now that's not surprising, it's very tricky. So we started encasing them, and talking to ink makers, and paint makers, surface coatings people, you know, "...here's some encased chemicals. See if you can make some inks and some surface coatings..." This was pretty hard work, because most of them still couldn't use the encased chemicals. Basically what they were doing was taking them, sticking them into an existing system, and hoping that would make the product. Now that would never work, and it never did work. So we employed ink specialists, and we learned how to make inks, so we could basically do it ourselves".

Having developed its range of inks, Multichem began looking for consumer applications. The Business Development Manager describes the sense of anticipation at that time. "It was really a very exciting time because we had these [inks] which people were able to use, just about, with a lot of technical assistance, but they could do it ".

The general understanding was that only volume markets were worthwhile. Because of the problems of handling specialist chemicals, even in capsule form, niche markets would have demanded disproportionate customer support. "It's too difficult to do the smaller markets. Each market you sold the specialist chemicals to required a large amount of technical back-up, due to the difficulties working with them. Working with a lot of smaller markets would have required too much technical assistance on the part of the company. The markets which were considered were all large enough markets to justify the company undertaking the products".

There follows an outline account of one of the NPD programmes through which Multichem aimed to find a market for its inks. There were other programmes that followed a broadly similar pattern, but it is not possible to present details within the confines of this conference paper.

The case study product: printed clothing

The idea behind the clothing project was to print garments with the special inks. Clothing manufacturers were contacted by the marketing department to obtain an initial reaction to this product concept. This was favourable, and so the decision was made to go ahead. This immediately set the agenda for an R&D programme: the further development of the specialist chemical inks so that they would be suitable for screen printing onto textiles.

Problems included: washability, "...the water resistance was generally atrocious, after five washes you'd basically got rid of it . . . so we had to do a lot of research into how we could make these screen printing inks stable enough to put onto fabric. We spent a lot of time with cross linking systems to try and get it washable;" and the loose weave of cheap mass market clothing fabrics, "the thing about [these materials] is that they are generally very low quality, very stretchy ... with lots of holes in. And we had an ink which didn't stretch and didn't wash very well ... also the [brightness of the] effect depends very much on getting a single layer of the chemical on a black background. If you've got holes everywhere, you don't get this single layer, ... and you don't get the pure effect;" and the contamination of the specialist chemicals by the other ingredients of screen-printing inks, "... the ink system we have had its limitations, you can't just stick your normal additives into the ink system, you can't use the normal binder systems ... so you're very limited by the ingredients you can put in;" and, for a cheap, mass-produced clothing, the cost of the specialist chemical-based inks was a chronic anxiety, "it's expensive, and ... most screen printers fell off their chairs when you told them that's what price they're looking at".

In spite of all these problems, the project pressed ahead: "So the market situation was wrong, there were [competitive products] at the time, the cost was too much, and there were technical problems as well. But we weren't to be stopped on that one ... we were not to be deterred, basically the size of the market was very attractive. We spent more on R&D...".

To some extent the problems of washability and of printing onto stretchy fabrics were solved. "It was really very close to what we needed. You still only got ten washes out of it, and they all had to be hand washed. You couldn't stick it in a machine. So it was getting a bit dodgy for the volume market, but people liked it".

In the light of the continued high cost, however, the limited washability which had been achieved was beginning to look insufficient: "If it's a very cheap sort of material, then OK, ten washes is fine, but with it being a specialist chemical product it wasn't cheap ... we were really having to charge quite a lot of money".

Meanwhile, the progress towards an achievable product on the technical front appeared to have the effect of clarifying the customers' requirements: "at that time, I did a lot of running around talking to the market ... When you talk to people about [this product] they get really excited, saying it's the most wonderful thing ever, but when we actually came to do it, they all sort of went a bit lukewarm. You can describe [the effect] on the phone, and they get very excited about it, but when they actually see the end product ... it's not like they'd been expecting. So there was always that disappointment factor".

What finally killed the project was competition from a rival technology. Not anticipated at the launch of the specialist chemical development programme, these dyes turned out to be cheaper, easier to apply and gave a better visual effect. "...That was another little nail in the coffin for us. It's basically a different type of material... And they are easier to use generally... I wouldn't say they are any more stable, they're just [give a better effect], easier to use and cheaper...so they came at just exactly the same time as we were trying to make people interested in [our product]. And they came in with all theirs at a low price. And we didn't even get a look in to be honest".

Chemtech

Although it is a small company, with only 60 or so employees worldwide, Chemtech is one of the industry's successes. Current worldwide annual turnover is about $10 million, increasing at about 20% per annum. The UK operation was set up about 5 years ago and currently employs 15 people, many of whom have worked for Multichem at some point in their careers. Though the company has interests in other technologies, about half of the UK employees are directly involved in specialist chemicals. The form of this involvement is rather different to that typical in Multichem. Chemtech neither develops nor manufactures specialist chemicals on its own account. Most of its raw materials, in fact, are purchased from 3 or 4 chemical manufacturers including Multichem. Rather, it is a company that has built up a particular expertise in formulating specialist chemical mixtures which will deliver the properties required in consumer applications. Surprisingly perhaps, in a science-based industry, this expertise has been acquired through experience, rather than formal scientific study.

Our data are from interviews, totalling ten hours, with the company's research, marketing and production staff. The quotations are from taped interviews with the UK Director: "the industry [is] founded really on empiricism. People go out and find a bunch of chemicals, and then mix them or process them in some way or another, and the ones that work you keep, and the ones that don't you throw away. So if you go and search the patent literature or the open literature there isn't very much written about anything that anyone does in the industry. There are patents there if you read them [that] tell you some rudiments, but there's nothing of any significance if you were to look seriously at getting into the industry. So there's quite a significant barrier to entry based on the high level of empiricism ".

This means that Chemtech's ability to adapt specialist chemical mixes is relatively unique within the industry: "... People that are in the industry who compete with us basically rely on their raw material suppliers to provide them with technical support. None of them really have the capability to develop anything that is outside of the ordinary".

The two basic processes through which Chemtech has built up its portfolio of capabilities are acquisition and recruitment. The following examples are typical. "We added a downstream processing capability in 1982 when we hired a guy from [one company] which was a competitor ... We also hired another guy who was doing [work for] medical [applications]. So he had some experience of novel specialist chemical formulations when he came to Chemtech".

Though the company is well aware of the importance of marketing the capabilities built up in this fashion, the industry as a whole has a reputation for uncertain quality. For this reason the company's ability to develop specialist chemical applications is best publicised by results and by the education of potential customers rather than advertising as such: "We don't do a lot of advertising ... you've got to go through an education process to build an new image for the products and the markets ... a lot of our competitors actually make products where the quality actually isn't what it should be, so you get a raw deal in terms of the image in the marketplace. Particularly in the UK market, the consumer level [of expectation] is a result of our efforts to ensure the quality of the products that we supply".

A very important consequence of the reputation built up by Chemtech is that product concepts normally originate with its customers rather than the company itself: *Interviewer*: "So what's the process then? Does someone come to you with an idea, or do you have an idea for a new product?" *Interviewee*: "No usually people come to us with ideas. We've got a good track record of making big things happen".

The case-study product: toy bricks

This happened in the case of a multinational producer of toy building bricks. The idea was to include a part that used the special effects produced by these chemicals in one of their toy sets. Chemtech became involved as a result of the toy manufacturer's search for a partner who could develop such products: "large companies undertake global searches for people within this business ... and they scan through the various trade publications, publications about company services and capabilities, and they'll end up with a shortlist of about a dozen people that might be worth contacting around the world, who appear to know something about the technology".

In its initial approach, the toy company had in mind a fairly definite specification for the brick. It is an important part of Chemtech's approach, however, that this was treated as a basis for negotiation rather than as an unalterable NPD target. The aim was to reach an acceptable compromise between the customer's requirements and what could realistically be achieved through the formulation of specialist chemical mixtures: "I think what they wanted was basically somebody who understood what they wanted to try and do. When we first set out to describe what was possible, we had to persuade them that they don't actually need the device to satisfy every one of their design prerequisites all at the same time. You end up working backwards from the ideal product to a product that: is going to be acceptable to them; that can be produced to meet their tolerances; is technically possible; and fits also their budget ... you have to communicate with [the customers] and let them understand the technology of your business. If they understand, you're not just fobbing them off with something that's second best, you're actually taking what's available and customising it to what they particularly want".

To Chemtech's director, this initial negotiation is a key phase of the NPD process: "customers usually ask us to do something which is close to being impossible, on the boundary between what you can and cannot do technically.

You've got to bring it back inside that boundary as far as you can so that you can give yourself a realistic shot of producing the thing reproducibly, and hitting it, not spot on every time, but certainly within the levels of tolerance that are acceptable to the customer. Everything that Chemtech has ever done with big multi-nationals has always involved some kind of dialogue with the customers. You listen to what they tell you to find the best shot you can give them".

Because the negotiation, on Chemtech's side, is based on an assessment of technological feasibility, it also serves as a vetting mechanism. The company does not involve itself in speculative R&D programmes: "...while you take the company forward, you have to turn your back on a lot of things as well. One of the reasons Chemtech has been successful is that we turned our back on more projects than we decided to pursue". *Interviewer*: "So what factors influence your decision about whether or not to follow an idea?"*Interviewer*: "Well I think a lot of it's gut feeling . . . the company depends a lot on me to make the judgement calls on what's worth chasing and what isn't worth chasing".

At the time of our research, the toy project looks like a success for Chemtech. This is not only important in itself: it will also feed back into the company's reputation-based approach to marketing: "this [toy company] thing is a prestigious piece of work. It's just [these chemicals] on a unit, a little widget which produces a certain effect. It's not that sophisticated a product really when you look at it, but it's made to a very demanding quality. Our ISO 9000 qualification was an important factor in taking the project forward. The new product's going to be launched in July, well we've made a few shipments, but when that appears on the market it will open people's eyes to the fact that high spec devices can be made to satisfy the most demanding clients using [this] technology. That's one of the things which our competitors around the world would find very difficult to do".

DISCUSSION

The comparison is of NPD chains, not companies

It is tempting, at this stage, to make a straightforward comparison between the relatively successful NPD process at Chemtech with the instance of failure at Multichem. Clearly the processes in the two companies differ in the source of the product concepts, the preliminary evaluations of their prospects and the monitoring of competing technologies. On the basis of such a comparison, it would not be difficult to put forward explanations of the companies' different success rates, pointing towards the conclusion that companies competing in fields characterised by rapidly evolving technologies ought to do it like Chemtech and not like Multichem.

The difficulty with this apparently logical approach is that Multichem and Chemtech are not altogether independent. Although they compete as the suppliers of specialist chemicals for new product applications, those supplied by Chemtech are the end-point of a secondary development based on chemicals supplied by manufacturers - including Multichem. Whilst Chemtech is not locked into this arrangement, nor even into this particular technology as such, its present reputation as a developer of specialist chemicals for consumer applications depends upon the existence of a supplier of 'generic' specialist chemicals. In this sense, the problems of Multichem - at least insofar as they stem from its situation as a supplier company rather than from strategic 'errors' - are part of the cost of Chemtech's relative success.

For this reason, we believe is better to view the NPD processes within these two firms as part of larger chains of relationships connecting the development and production of specialist chemicals on the one hand with the consumer end-point on the other. From this point of view, the case studies are windows into two different product development networks, and it is the effectiveness of these as a whole that is in question, rather than that part of the NPD process which takes place in the two firms.

Searching for mass vs. niche markets

In pattern 1, product concepts originated with the specialist chemical manufacturer. Since this was a fairly large company, this meant that they were driven by the need to find outlets for the company's production facility. In Multichem, this requirement was translated into a scan of (existing) mass-produced products for those that might be enhanced by the chemicals' properties. Unless the properties in question were unique, therefore, this strategy virtually guaranteed that the company would encounter competition from other technologies. If some feature of a mass product is obviously attractive, it is more than likely that that others will be looking for ways of providing it.

In pattern 2, by contrast, the product concepts suggested to Chemtech by its potential customers might well be have appeared to Multichem as the kind of 'novelties' to be ignored in the search for mass markets (the toy brick was fairly typical in this respect). The fact that the market for some of these 'novelties' later expanded to several million units annually, could scarcely have been predicted at the product concept stage - which means that the search for mass markets built into the concept stage of Multichem's NPD process is virtually guaranteed to overlook such opportunities. To some extent, this was a function of the scale of the companies: since Chemtech could prosper in markets too small to support Multichem, it could not only develop products intended for these markets, but also reap the rewards if and when those markets expanded.

A possible (preferred?) solution for Multichem might have been to withdraw from the attempt to develop consumer products directly and to concentrate on supplying specialist chemicals to companies like Chemtech. Multichem was aware of the attractions of this strategy, but the problem was that there were simply not enough companies with Chemtech's capabilities.

Vetting projects

Vetting, in pattern 1, consisted of two phases. The first was a preliminary canvass of the product concept amongst potential customers. Where this was positive, the company proceeded to the 'working model' stage of product development, thus guaranteeing that the costs of this would be incurred every time a customer was attracted to the concept. The second phase of vetting was tacit in the customer's trials of the performance and manufacturing compatibilities of the prototype product. In the cases outlined above, NPD on pattern 1 was abandoned only at the point where the product failed these trials, having incurred all of the costs of development work up to that point. Thus vetting in pattern 1 is, effectively speaking, in the hands of the customer. The anticipated difficulties of developing appropriate chemicals appear to have been of little account. They appeared, indeed, to have been regarded as a scientific challenge, of which more in a moment.

In pattern 2, many possibilities were rejected at the concept stage. Chemtech would only embark on NPD on the basis of a 'gut feeling' (i.e., experience-based knowledge of what can be achieved by mixing the basic chemicals) that the development track would be short and successful. In this sense the company was conscious of operating at the 'low tech end of a high-tech field.' Given that the initial approach in pattern 2 was from customers, and that this signified that that a potential market existed, this meant that vetting was on the basis of technological viability, that it was carried out by the producer company, and that it mostly took place before development costs were incurred.

Customer involvement and commitment

In pattern 1, although the marketing department was involved jointly with R&D in originating the product concepts, it was a marketing department dominated by research-minded chemists. This meant that the product concepts were the union of a highly professional knowledge of specialist chemical properties and a decidedly amateur understanding of the relevant markets. Though the company was aware of the importance of marketing, this awareness took the form of testing product concepts against the reactions of relevant consumer product manufacturers. Though the reactions themselves were probably authentic, the market sector from which they were obtained had already been defined by the initial 'R&D' view of the market. In this respect the formation of the product concept in the pattern 1 NPD process failed to draw on the marketing expertise of the consumer product manufacturers.

In pattern 2, the product concept originated with the consumer product manufacturer, rather than Chemtech. Since these manufacturers were large companies, the anticipated competitive advantages of a specialist chemical-based product were likely to be informed by a relatively sophisticated understanding of the market. Thus Chemtech could be fairly sure that a technically successful NPD process would also be commercially successful. Besides this advantage in marketing, surveys in the manufacturing, medical instrument and food processing industries consistently indicate that the use of the customer as a source of technical ideas is associated with NPD success (Germunden *et al.,* 1992; Shaw, 1993 and Hutcheson *et al.,* 1996).

As with customer involvement, so with commitment. In pattern 1, there was no customer commitment until relatively late in the NPD process, and not always much even then. The prototype specialist chemical application was developed unilaterally by Multichem on the basis of no more than an expression of interest in the concept on the part of the customer. Even when an achievable process had been demonstrated, the customer's involvement in developing it into a commercially viable proposition tended to be limited to setting the design specifications and to the running of manufacturing trials (the clothing case). This limited commitment on the part of potential customers was probably a function of Multichem's strategy of looking for applications in existing mass markets. Almost by definition, the production technologies in such markets will be highly developed and stable, so that any modification would incur considerable costs. As a result, there appeared to be no possibility of a compromise between the consumer product specification and the properties of specialist chemicals. Noticeably, in fact, the requirements of Multichem's customers became clearer, and more demanding, as development work proceeded.

In contrast, Chemtech's director was keenly aware that an approach from, rather than to, a customer signalled commitment to the success of a project: "... You know, people's careers are made on bringing the products we are able to produce for people to market successfully. I mean look at [an electronics-based manufacturer], the first time round they failed to bring the thing to market they ended up firing a few of their research engineers. And I think this is happening with [one company] at the minute in the cosmetics industry. These high volume, high exposure deals are very important to the people that we're dealing with in the company as our customers, and they can't afford to drop the ball, because their careers are on the line".

Shaw (1993) has noted the importance of product champions in customer companies in facilitating learning by innovating entrepreneurs and in diffusing the costs of NPD programmes. In the case of Chemtech, this customer commitment also meant that their customers were more likely than Multichem's to be flexible on performance specifications. For this reason, Chemtech was able to negotiate the initial targets for the NPD process, and to continue the dialogue "from first contact to first order", as the Director put it.

Doing science vs. doing new products

In pattern 1, NPD was driven by the R&D mentality. This was probably a function of the history of Multichem, as well as the background of its staff. Much of the initial growth of the company came from supplying scientific apparatus to universities and colleges, and this was still an important side of the business. Appropriately - from the point of view of this side of the business - many of the staff, including the marketing staff, had higher degrees in chemistry. What ran like a thread through the Multichem interview was a relish for challenging R&D problems. Instead of prompting re-evaluations of the NPD programme in which they occur, it seems that the difficulties only added to the company's determination. Entirely appropriate to an R&D team tasked with the development of a new technology, such an approach is more questionable when carried over into an NPD strategy. It is an attitude guarantees that expenditure on each NPD programme will end only with its complete success or complete failure.

In pattern 2, NPD was driven by the intersection between the customer's product concept and Chemtech's expertise in mixing specialist chemicals. The director of Chemtech was conscious that the firm's approach lacks the glamour of 'real' science. NPD tracks appeared to be short, highly specific to the customer's product (rather than aimed at a generic mass market), and probably inexpensive.

In the long run, of course, the commercial successes of Chemtech within pattern 2 could be argued to be dependent upon the scientific knowledge of specialist chemical properties produced within pattern 1. Amongst other things, this was signified by the fact that many of Chemtech's staff had previously worked for Multichem. From this point of view, the problem might be seen as one of the distribution of the fruits of success within pattern 2, rather than the failures of pattern 1.

Establishing the initial reputation

The major unknown in our case studies, as they stand, is the prior, and very large question, of how Chemtech created and maintained the reputation that attracted the initial approaches from customers. The next phase of our research will take up this question. For now, we can note that much of Chemtech's competitive advantage was built upon its reputation for delivering chemical mixes to suit customers' requirements. As one of only two UK companies recognized as possessing this capability, Chemtech's position is very strong.

CONCLUSIONS

Evidence from case studies is, in its nature, indicative rather than conclusive. The successful and unsuccessful NPD programmes compared this paper, however, point towards two main conclusions.

The first is that companies with a successful record in the development of a technology may not be equally successful in the development of products based on that technology. In order to be successful, NPD programmes need to aim at a new, or markedly advantageous, product, as experienced by the user (Hutcheson *et al.,* 1995). Product innovation in this sense is not guaranteed by the fact that a product application involves the solution of difficult technological problems. The mindset and company culture adapted to technological innovation may tend to under-emphasise the product innovation aspect.

The case in which technological development and product innovation were successfully integrated occurred through the agency of an intermediary company. This company had built up an experience-based expertise in the adaptation of the producer-company's technology to consumer needs. Because of the intermediary company's credibility in this field, the product concepts underlying its NPD programmes originated with the manufacturers of consumer items, thus ensuring that the product innovation aspect had been given due weight. Thus a second conclusion might be that a network in which the technology producer connects to consumer product manufacturers through an intermediary company specialising in applications of the technology might be more effective than a direct connection. That this differential success was not confined to the particular NPD programmes that we studied was indicated by the relative financial status of the producer and intermediary companies. Whereas the producer company was near insolvent, the intermediary company was profitable and expanding rapidly.

In this, and similar instances, however, it is important to bear in mind that the success of the intermediary company was based, in part, on the technology obtained from the less profitable supplier company. Whilst it is possible for individual intermediaries in this position to reduce their dependence on particular technologies by diversifying into other capabilities (as had happened in our case study), this does not diminish their dependence as a class on the suppliers of technology as a class. If, therefore, intermediary application companies are to be encouraged as a means of integrating product and technological development in fields of advancing technology, some thought also needs to be given to the redistribution of the rewards, from these application companies to the suppliers of the base technology.

REFERENCES

Bower D.J., 1993, New product development in the pharmaceutical industry: Pooling network resources. In *Journal of Product Innovation Management,* Vol. 10, No. 5, (New York: Elsevier Science),pp. 367- 375.

Cooper, R.G., 1988, Predevelopment activities determine new product success. In *Industrial Marketing Management,* Vol.17, pp. 237-247.

Gemunden, H.G., Heydebreck, P. and Herden, R., 1992, Technological Interweavement - a Means of Achieving Innovation Success. In *R & D Management,* Vol.22, No.4, pp.359-376.

Hutcheson, P., Pearson, A.W. and Ball, D.F., 1995 Innovation in process plant: a case study of ethylene. In *Journal of Product Innovation Management,* Vol.12, No.5, (New York: Elsevier Science), pp.415- 430.

Hutcheson, P., Pearson, A.W. and Ball, D.F., 1996, Sources of Technical Innovation in the Network of Companies Providing Chemical Process Plant and Equipment. In *Research Policy*, 25: pp. 25-41.

Karakaya, F. and Kobu, B., 1994, New Product Development Process - an Investigation of Success and Failure in High-Technology and Non-High-Technology Firms. In *Journal of Business Venturing,* Vol. 9, No.1, pp.49-66.

Moenaert, R.K. and Caeldries, F., 1996, Architectural Redesign, Interpersonal Communication and Learning in R&D. In *Journal of Product Innovation Management,* 13, (New York: Elsevier Science), pp. 296-310.

Noren, L., Norrgren, F. and Trygg, L., 1995, Product Development in Interorganizational Networks. In *International Journal of Technology Management,* pp.105-118, pp. 41-55.

Shaw, B., 1993, Formal and Informal Networks. In *The UK Medical Equipment Industry Technovation,* Vol.13, No.6, pp.349-365.

Tidd J., 1995, Development of Novel Products Through Intraorganizational and Interorganizational Networks - the case of home automation. In *Journal of Product Innovation Management.* (New York: Elsevier Science), Vol.12, pp. 307-322.

Acknowledgement: This paper was written in the course of a research project directed by Anne Tomes and sponsored by the UK Design Council on the role of the design imagination in connecting basic research to product application. The arguments and conclusions presented in the paper, however, are entirely the authors' own.

An Innovative Approach to Developing the New British Standard on Innovation Management

Alan Topalian and Bill Hollins

This paper outlines the process by which British Standards are prepared and provides a brief history of design management standards. It also touches on the challenge of how to increase dramatically the number of companies that adopt British Standards in their operations. The proposed radical new BS 7000: Part 1 on managing innovation is then discussed which will provide guidance on how top executives, their designers and other creative specialists might plan to manage the design of products required more than ten years in the future. But is it possible to predict far enough into the future for a standard to be written? Can one set guidelines for an inherently 'messy' process like innovation? This paper will explore how designers and their clients can work to longer time horizons so as to create designs that compete well into the future.

"One day, Prime Minister Margaret Thatcher asked how come the Japanese were kicking the hell out of UK industry? She was informed that their quality was better than ours. Five years later she asked how the Japanese could still be kicking the hell out of us; she was informed they had better design. Another five years on, the Japanese continue to kick the shit out of us, and the reason is that they produce the right products and we don't". Professor Tony Stevens, Loughborough University of Technology, 1994.

THE BRITISH STANDARDS INSTITUTION

The British Standards Institution (BSI) is the organisation that produces national standards for use in the United Kingdom. British Standards are also used in other parts of the world, especially Commonwealth countries (such as Australia, Canada, India and many parts of Africa) and Scandinavia.

The BSI also represents the United Kingdom's interests in the development of European (EN) and international (ISO) standards. A 'standard' — be it ISO/IEC, CEN/CENELEC or BSI — is defined as "a document, established by consensus and approved by a recognised body, that provides, for common and repeated use, rules, guidelines or characteristics for activities or their results, aimed at the achievement of the optimum degree of order in a given context".

HOW STANDARDS ARE PRODUCED

Proposals to create new standards are presented in writing to the appropriate BSI technical committee as a *'New Work Item'* with information supplied under a common set of headings. Approval to proceed is given by the Sector Board that manages a group of technical committees. It must be satisfied that work is likely to result in a worthwhile standard within a sensible time. A budget is set though those directly involved in committee work are not paid (other than committee secretaries who are BSI officers).

Once approval is granted, the technical committee usually convenes a small drafting panel of experts to compile the standard. Committees are open so recognised organisations that wish to contribute may take part by nominating representatives; a couple of suitably qualified individuals may also be co-opted. This means that committees frequently include members who have vested interests through their organisations and tend to promote these interests.

British Standards are written around consensus. This is often a painfully slow process as drafting panels meet on a monthly basis and it is seldom easy to reach a consensus. Care must be taken so that what is agreed does not lead to a bland document offering vague guidance. The trend is to make standards (and particularly guides) more user friendly; requirements standards are much more prescriptive hence, some argue, 'drier'. The use of devices such as flow diagrams and practical checklists is encouraged to provide readers with quick guides to areas that need consideration. When a draft is complete, the panel submits it to the main committee that decides whether to accept it or request revisions (almost certain); on rare occasions, a draft may be rejected completely.

Accepted documents become 'drafts for public comment' which are issued for a period of three months and anyone can purchase copies from the BSI. During that time, the BSI accepts comments, suggested amendments and developments from whoever cares to write in. Individuals may also object to or encourage publication of those standards. All this input is considered in further meetings of the panel and the revised draft incorporating agreed changes is resubmitted to the technical committee. If satisfied that the feedback received has been taken properly into account, the committee votes to approve publication of the standard.

Before publication, draft standards are appraised by BSI's editorial staff to ensure copy style, structure and presentation comply with BS 0, *A Standard for Standards* (BSO, 1997). This may lead to further 'negotiation' with the committee. When all parties reach agreement, the standard is issued. Typically, the process from *'New Work Item'* to publication takes three or so years.

MARKETING AND ADOPTION OF BRITISH STANDARDS

After undergoing lengthy drafting and consultative processes, British Standards rarely get wide exposure. Few are launched formally at promotional events, and there is virtually no advertising in 'outside' publications. Consequently, sales may be minimal because potential users don't know of them.

The BSI is upgrading its marketing approach but, for historic reasons, it may be some time before it evolves a truly pro-active stance. To date, sales volume has not been a primary consideration and it should be recognised that a significant proportion of standards will not be sold in large numbers. They are actually written for the public good and the benefit of British industry even though the market may be tiny. For example, it is essential to have standards covering petrol though these will probably be sold to a relatively small number of oil companies and related suppliers; car drivers are unlikely to purchase copies. The same applies to most standards concerning safety. 'Guides' are more likely to sell in larger quantities. Second, the BSI has relied heavily on interested bodies represented on their committees to promote new standards among their members and so gain wide publicity. Unfortunately, this does not always happen.

BS 7000 – THE WORLD'S FIRST DESIGN MANAGEMENT STANDARD

Work on the world's first design management standard — BS 7000 *Guide to Managing Product Design* (1989) — began in 1985 with a committee of 26 individuals drawn from all areas of design spanning from the armed forces to fashion design. Very little progress was made after a year of monthly meetings, partly because it became obvious that those round the table understood 'design' to have different, even contradictory, meanings.

The standard was set out in four principal sections: an Introduction led to a section on "managing design at the corporate level", followed by a section on "managing design at the project level". The standard concluded with a section on "managing the design activity". At each level, design management is discussed under four sub-headings: planning, communicating, monitoring and control. The standard was twenty pages long and the fact that it took four years to write gives some indication of the disagreements that occurred during development.

A key issue was whether the standard should be a 'specification' or a 'guide'. Organizations can be instructed to conform with and be assessed against the former — BS EN ISO 9000 is such an example (1994). Several committee members had a vested interest in creating a new specification standard, hoping to become assessors of its use. However, there was insufficient knowledge on 'best practice' in design management to insist that organisations work to set guidelines. So it was decided to issue the standard as a 'guide' which proved to be the right decision. One example that underlines this was a diagram that depicted an 'idealised design process' with 'detail design' followed by 'design for manufacture'. Had this been a specification standard used to accredit user enterprises, such a recommendation would have decreed against the subsequent adoption of Concurrent Engineering.

BS 7000 was published at the end of 1989 and went on to become a 'best seller' in BSI terms with sales of between 100,000 and 200,000. Unfortunately, figures for the early years, when sales were highest, are no longer available because sales records were held on a three-year rolling basis. As standards are normally updated every five years, it would be useful for sales records to be kept for at least five years and released quarterly to relevant committees.

ENRICHING THE LANGUAGE OF DESIGN MANAGEMENT

Early on in the drafting of BS 7000, the committee recognized that discussions on design management were handicapped by the confusion about even basic terms and the paucity of language. These, in turn, had a detrimental effect on the development of the discipline. This led to the formation of another panel to clarify the definitions of core terms. The panel subsequently broke with tradition by adding several new terms, particularly relating to the corporate level, thus enriching design management language. A new standard BS 7000: Part 10 *Glossary of terms used in design management* was published in April 1995. Unfortunately, this appears to be largely ignored to the detriment of design practitioners and their clients. Clearly, it is difficult to raise performance when terms are used in misleading ways thus complicating communication and limiting serious debate.

The committee also felt that a guide to the preparation of specifications would be an additional help. BS 7373 *Guide to the Preparation of Specifications* was duly published in 1991 and an update has just been completed for publication in late 1998.

COMPLEMENTARY SECTOR SPECIFIC STANDARDS

One of the criticisms of the original standard was that it took too broad a view and was insufficiently focused on how design might be managed in different industry sectors. Other 'stand alone' standards were proposed, derived from the original, aimed at these different audiences to facilitate wider adoption.

Work on these complementary standards started in summer 1992 and the number of the original standard was altered to BS 7000: Part 1. To date, the following sector specific standards have been published:
- Part 2 Guide to Managing the Design of Manufactured Products (1997).
- Part 3 Guide to Managing Service Design (1994).
- Part 4 Guide to Managing Design in Construction (1996).

Understanding of design management has advanced since the original BS 7000 was published. Whilst the original publication is still instructive for organisations new to the discipline, others have progressed beyond many of its recommendations. The newer standards — particularly Part 2 — incorporate more of this later thinking and are structured differently.

In these standards, 'design' is taken to encompass a 'total process' from conception through to disposal. This is generally known as "Total Design" (Hollins and Hollins, 1991; Pugh, 1991). Particular attention is paid to the early, low cost stages of the management process when most of the expenditure is committed (though it is actually spent in later stages).

A basic, general process (or model) is defined which provides the 'backbone' of the guide. Another difference from the original BS 7000 relates to the fourth section entitled "Managing the Design Activity". When drafting Part 3, the committee found that the original structure did not apply well to small firms, not least because of the extensive overlap between "design project management" and "managing the design activity" in those enterprises. So in Part 3 and Part 2, the third section provides an overview of managing the design process while the final section goes into this in greater detail with several helpful checklists.

INVESTIGATING HOW STANDARDS CAN BE USED MORE EFFECTIVELY TO CREATE COMPETITIVE ADVANTAGE

To ensure that industry gets the standards it wants, it is necessary to find out how standards are viewed by potential users. This is critical as standards are not cheap (typically £40 upwards) and sets of standards can involve significant financial outlays.

Little information is available on how different business enterprises find out about, respond to and go about adopting British Standards. Evidence suggests that awareness of design management standards is relatively low, consequently take-up is poor and only a small proportion of organisations benefit from their guidance.

In 1996, the Chartered Society of Designers, the Institution of Engineering Designers and the BSI submitted a joint bid for funding from the Design Council in the United Kingdom for a survey to fill that surprising gap in knowledge. Their objectives were to:

- Investigate how different business enterprises find out about British Standards relating to the management of design.
- Determine the channels of communication that reach potential users more effectively.
- Map out the process by which enterprises respond to and use standards.
- Ascertain the reasons behind the decisions to use (or not use) standards, the problems encountered during their use, and reasons for dropping standards.
- Draw lessons in order to enhance awareness of such standards and facilitate adoption.
- Help business enterprises to make most effective use of standards.
- Identify how format and presentation can be made more attractive and productive.
- Improve the marketing of BSI design management standards.

Particular attention was to be paid to small and medium-size enterprises (SMEs), especially those involved in markets where there is significant competition from imports and those that export products and services. It was hoped that the project would result in greater understanding of the role of standards in promoting profitable growth in SMEs as well as the formulation of guidelines on best practice in adopting new standards to gain greatest benefit.

Up to 250 target organisations across the United Kingdom were to be surveyed by questionnaire and one-to-one interviews, including enterprises that had recently purchased design standards (already interested parties) and those involved in product creation but were not using standards. Responses to BS 7000: Part 2 which was to be launched in March 1997 were to be monitored, thus providing a 'live' case. A further dozen case histories of particularly instructive experiences were to be written up.

The Design Council did not back the bid on the grounds that such research was already under way. No information has been available yet on this research though, hopefully, results will be released as soon as possible to those who need them, especially the BSI.

PROPOSED NEW BRITISH STANDARD ON INNOVATION MANAGEMENT (BS 7000 : PART 1)

All the current standards in the BS 7000 series provide guidance to those who know what they want to design in the short- to medium-term. The original BS 7000: Part 1 became obsolete when Part 2 was published and was withdrawn in April 1997. It was realised that future work could proceed in several directions, however the priority was to introduce an 'umbrella' standard as the new Part 1. But what should be the subject matter of that standard?

A major criticism of the original BS 7000 was its failure to look far enough ahead and address the formulation and implementation of long-term design management strategy. So it was proposed in January 1995 that the new Part 1 should take on the challenge to help organisations to plan winning products two to three generations hence — in essence, a guide to the management of innovation and competitiveness for those who do not know what products / services they should be designing to sustain profitable ranges 10 to 20 years from now.

The new BS 7000: Part 1 is conceived as a radically different standard to help UK plc get ahead and create more world-beating products, hence more world leading enterprises. It is a bold attempt to break new ground as well as develop the standard in an innovative way. The document will have a corporate, as opposed to a project, focus with top management as its prime target audience because the quest for greater creativity and innovation should start with, and be driven by, board directors and owner / managers.

The guide will major on the 'design dimension' of strategic thinking and planning of business futures. It will address future strategy in more detail and guide users in how to assess and harness their corporate capabilities, generate market intelligence and create the right organization to benefit from this information. Guidance may be given on the following:

- Understanding innovation (different types of innovation, innovation process, inherent temporariness of innovative achievements, need to move on swiftly, etc.).
- Key dimensions / issues of innovation: the 'Innovation Universe'.
- Roles and responsibilities (especially at board level), and how to infiltrate 'innovation' elements into the job descriptions of all appropriate executives and staff.
- Planning for innovation / corporate 'refresher' campaigns for innovation.
- Clarifying the implications of innovative objectives, strategies, policies and action; strategic contribution of design; design in strategic planning; design leadership.
- Promote need for a more pronounced and unrelenting innovative stance (Encouraging diversity, out-of-bounds thinking and experimentation to generate rule-breaking innovation; degrees of 'newness'; technological / innovation monitoring).
- Establishing (and sustaining) an innovative culture and corporate process.
- Getting new practices going (and keeping them fresh and relevant); rehearsing the future through 'experience management'; exploring the implications.
- Anticipating and organising to avoid criticism and wrecking tactics from those who oppose change.

- Introducing / launching innovations.
- Communications strategies to explain bases of innovations within / outside host organisation.
- Skills acquisition and development; rehearsing the experience of going over the 'innovation barrier'; fast application of new knowledge and skills to respond to / anticipate changes in the environment.
- Costs of innovation, returns on investments; principal components making up innovation budgets.
- Funding innovation.
- Alliances that boost innovative performance.

PIONEERING A NEW APPROACH TO STANDARDS DEVELOPMENT

Unsurprisingly, the committee has struggled with the issues that should be addressed and how to break out of the mould of traditional standards in terms of content and presentation. An initial trawl for suggestions of issues and approaches to cover in the new standard drew few novel responses and indicated that executives in industry have great difficulty viewing the development of products beyond their next generation. Tools / techniques such as market research, benchmarking and business process re-engineering, whilst important, are more appropriate for products being developed for current circumstances rather than ten years in the future.

So the committee resolved to approach innovative organisations in the hope of establishing collaborative relationships to share ideas and explore issues together for mutual benefit. This lead to a 'brainstorm' session hosted by the Director of Innovation of a major industrial group that brought together five executives (mostly board directors of group companies) and six BSI committee members.

It is hoped that the BSI will shortly launch a survey into how innovative enterprises adopt effective long-term strategic perspectives. Thirty companies considered to have evolved such successful perspectives in new product development will be questioned to see whether any patterns emerge or recommendations on 'best practice' might be distilled to benefit others.

Design models are vital for the present and fine for, say, three-year time 'slices'; but are less effective when planning well into the future. No matter how much iteration is built in, their relative inflexibility and short-term horizons severely limit their value. What is needed is more of a holistic, systems approach.

A good starting point would be to look at organisations currently working in areas that are changing fast. Though it is widely reported that change is occurring at a faster rate than ever before, this is misleading. Just look around your room: how many items did not exist, and could not have existed, twenty years ago? Probably not that many. There may be a fair deal of relatively superficial change, but little in the way of solid step transformations. For example, the design of cars changed far more in the ten years from 1895 to 1905 than in the last ten years. Furthermore, people do not always welcome radical change in the products and services they use; they like improvement, but fundamental change can be very inconvenient, disorientating and costly. Little is undergoing radical change at present, though electronics, communications, pharmaceuticals and biotechnology are exceptions. Organisations working in these 'fast track' fields must have some of the keys to coping with radical change. Those that seek to survive long term should look at what these organisations do and emulate them as best they can. So what patterns do these fast track organisations have in common? They make the present pay so that they can survive into the future; they invest in long-term research funded by the present.

We often hear about 3M's '15% rule' whereby employees can spend 15% of their time on 'blue-sky thinking' and pursuing highly speculative ideas. *Post-It Notes* are often cited as an example of the fruits of this regime. What is overlooked is that the bulk of employees' time is spent doing standard product improvement to sustain the company now and pay for high-risk projects that might generate spectacular profits in the more distant future.

The distinctive competence is the ability to 'switch on' those far-reaching searchlights and to harness creative energy to bring this exciting future nearer the present. So the challenge is to achieve the twin-track performance because a profitable present without a long-term innovation strategy is limited; similarly the quest to innovate over the long-term without profitable current operations is virtually impossible unless you have a fairy godmother to fund you indefinitely.

By viewing potential ventures within a strategic context, it should be possible to identify a series of parameters or 'boundaries' that will show the types of project that are right for a specific organisation. This will help people to identify those achievable areas worth pursuing as well as those areas that should be avoided. These frameworks provide focus and design direction channelling the designers' efforts towards options with greater chances of success. Often, success through innovation has more to do with timing and the manner in which innovations are introduced to market rather than the quality and execution of ideas.

LEARNING FROM EXPERIENCE TO LEAP FORWARD FROM THE PRESENT

20/20 vision afforded by hindsight is a marvellous asset. Yet many of us fail to harness the learning experiences offered by hindsight to increase our chances of future success (Topalian and Stoddard, 1997; Wassermann and Moggridge, 1990). We seldom bother to assess what might be done differently if we could have our time again. So, to rephrase George Santayana, those who fail to learn from the past are condemned to repeat it. Such repetitions lead to inordinate time and resources being wasted on fire-fighting and re-working old territory unnecessarily. Apart from dramatically reducing the prospects of success, we constrain ourselves to incremental development at best. Though dissatisfaction with the status quo can act as a drive to unlock creativity, irritation and frustration more often wear away our enthusiasm and blind us to fresh insights; they stunt our ability to break loose and create radically different futures — significant leaps forward are so unlikely as to be psychologically 'out of bounds'.

All of us can create opportunities to be better prepared even for events that are totally unimagined and unexpected. As Gary Player said, "The more I practise, the luckier I get". We, too, can practise to hone our approaches beforehand — even if only in our minds — as a means of familiarising ourselves with new and different situations, ironing out shortcomings and getting things as right as they can be first time round. We grow wiser, are more sensitive and open to new opportunities. We often become more tolerant to uncertainty and gain confidence to take greater risks. Consequently, we are able to move on with greater ease, and faster.

VISUALISING THE FUTURE BY MAPPING OUT THE 'CUSTOMER-PRODUCT EXPERIENCE CYCLE'

The future presents multiple possibilities — some beneficial, some threatening. To cope with these eventualities, we need to 'map' them out and formulate strategies to avoid or deal with them. That planning process and much of the subsequent actions are fundamentally design processes: we design desired futures then design the means to bring them about (Topalian, 1995). But dealing with the future innovatively may frequently feel like getting on a crowded moving escalator the wrong way: it is harder to make headway and you have to look further ahead to give yourself more time to cope with the future that is coming at you faster than you wish.

Visualisation is a key tool for 'previewing' those futures, typically through scenarios and concepts for specific products, services and environments. These enable us to gain a more tangible hold on 'what might be'. That extra tangibility then facilitates implementation. The facility to 'create the future today' is a characteristic of design leaders.

The traditional view of product design is the 'crafting' of physical entities. Though these are brought into existence to satisfy defined needs — in many instances, those of the creators rather than users — perceptions of satisfying needs tend to be limited to current use in defined circumstances. There is relatively little regard for how products come to be purchased, whether customers understand how to use them or exploit their full potential.

The marketplace is overcrowded with products of increasingly similar appearance and performance; their operation and the technologies they incorporate are progressively less understood by users. Customer satisfaction derives from the range of experiences from first awareness through to final disposal. Commercial success — for both industrial and consumer products — is achieved from managing those experiences: each phase has considerable potential to delight or antagonise potential customers and actual users. Understanding those experiences forms the soundest foundation to designing satisfying products so that valued 'highs' can be planned in and debilitating 'lows' eliminated as far as possible.

All such 'customer-product experience cycles' have common phases such as awareness, interest and information gathering, purchase, first use, on-going use and disposal. Each phase can be analysed in detail to map out key events and likely sequences. Rehearsing the acquisition, ownership and use experiences should help distil how and at what points design and innovation can facilitate and enhance those experiences. These rehearsals might take place in the design studio, a consumer clinic, retail outlet, order-taking office, supplier's workshop or user's premises. In this way, the impacts of behaviour and design decisions are explored on all parties together with the implications on key disciplines. Target audience reactions to product propositions can be evaluated to establish perceived value and priorities. The process by which they familiarise themselves with new products (especially 'out of box' and 'first use' experiences) and how they then go on using products can be similarly analysed. Helping customers to use products to their full potential may be another distinctive competence in 'the innovation game'.

Analyses of 'experience cycles' provide more powerful triggers for developing technologies that contribute to successful product design, manufacture and delivery. The dialogue that is established and nurtured between the core team, users, suppliers, distributors and so on, gets to the root of current perceptions, thinking and practices. Through growing trust and confidence, it also encourages exploration of the ideal future: the 'if only' world. Topalian and Stoddard (1997) contend that future products that match needs and aspirations closely, and deliver favourable customer experiences, are more likely to come from such analyses than from 'technology push' strategies. Use and ownership experiences provide very powerful targets to aim for in new product creation, especially when experienced personally by design / development team members. Moreover, the creation process will be more involving and inspiring because it is more human, fun and rewarding. When these experiences are enhanced through visualisation, you could not wish for a more vivid 'living' brief to work to. The design and innovation processes are effectively 'dialogues with the future'.

CONCLUSION

The mission statement of the British Standards Institution (1997) is " ... to increase UK competitive advantage, and to protect UK consumer interests". The BS 7000 series of standards goes right to the heart of this mission statement. Design is the integrating process that delivers satisfying, productive experiences such that all associated parties benefit. Therefore, it is increasingly recognized that one of the major ways to improve the competitiveness of UK plc. is through better management of product and service design.

Tomorrow's basic requirements are today's delight features (Deming, 1986). As such, it seems a sensible strategy for businesses to design for the future so as to excel in the present. The new BS 7000: Part 1 should help in this aim.

REFERENCES

BS 0, Revised. 1997, *A Standard for Standards.* (London: British Standards Institution).

BS 7000, 1989, *Guide to Managing Product Design (later Part 1).* (London: British Standards Institution).

BS 7000: Part 1, *(forthcoming) Guide to managing innovation and competitiveness.* (London: British Standards Institution).

BS 7000, Part 10., 1995, *Glossary of Terms Used In Design Management.* (London: British Standards Institution).

BS 7000, Part 2., 1997, *Guide to Managing the Design of Manufactured Products.* (London: British Standards Institution).

BS 7000, Part 3., 1994, *Guide to Managing Service Design.* (London: British Standards Institution).

BS 7000, Part 4., 1996, *Guide to Managing Design in Construction.* (London: British Standards Institution).

BS 7373, 1991, *Guide to the Preparation of Specifications.* (London: British Standards Institution).

BS EN ISO 9000, 1994, *Quality Management and Quality Assurance Standards* (London: British Standards Institution).

Deming, W. E., 1986, *Out of the Crisis.* (Cambridge: Cambridge University Press).

Hollins, G. and Hollins, W. J., 1991, *Total Design: Managing the Design Process in the Service Sector.* (London: Pitman).

Pugh, S., 1991, *Total Design.* (Addison Wesley).

Topalian, A. and Stoddard, J., 1997, 'New' R&D management: How clusternets, experience cycles and visualisation make more desirable futures come to life. In *Proceedings of Managing R&D into the 21st century Conference; Manchester July 1997*, Vol. 2.

Topalian, A., 1995, Design in strategic planning. *Proceedings of The Challenge of Change 3rd International Conference on Design Management,* University of Art and Design, Helsinki. August.

Wasserman, A. and Moggridge, B., 1990, Learning from experience: An Approach to Design Strategies for Product Success. In *Proceedings of Product Strategies for the '90s Conference,* (London: Financial Times), October.

Quantum Innovation: an open-systems approach to the new business of design

Nick Udall

This paper introduces an open-systems approach to the new business of design, whereby process, product and person influence each other in all directions. In this sense, New Product Innovation is a creative opportunity for an organisation to get to know itself, as well as others to get to know the organisation.

INTRODUCTION

The objective of successful business is not just to survive but to thrive. Depending upon the history, memory and experience of an organisation, it will sit somewhere on a meaningful continuum where survival (usually economic) is at one end and realisation of creative potential is at the other.

Organisations are simply groups of people who supply through product or service, other people. Organisations are run by people for people. An organisation of people can never (ontologically) be the same as another. Even if organisations were to share the exact same workforce, the agent or purpose which binds them together is its differentiation. This uniqueness gives an organisation the power to create. Its very existence relies on its ability to stand out from everyone else. Organisations are therefore in their essence meaningful, whether that manifests itself through the mediation of a physical, social, intellectual, emotional or spiritual offer.

Discovering and harnessing an organisation's uniqueness, by aligning the development of their people with the development of their products and services, is fast becoming a central strategy for ensuring survival and moving confidently towards living an organisation's shared aspirations. It is the premise of this paper that New Product Innovation (NPI) is a powerful and creative opportunity for enabling an organisation to get to know itself i.e., its creative uniqueness, as well as enabling others, in a more conventional sense, to get to know the organisation e.g., through customer centred branding, the invention of new and emerging markets, and shifting patterns of customer choice.

The aim of this paper is to point to some emerging principles and actions of the new business of design. The focus is on the systemic and developmental nature of NPI, whilst the reader is invited to close the ambiguous gap between theory and practice.

THE NEW BUSINESS OF BUSINESS

Innovation is the reproduction of creative ideas and inventions on a meaningful scale and within cost effective parameters (Senge, 1992). Innovation is undoubtedly central to business survival. Yet, the average life span of an organisation is only between 40 and 50 years (Kim, 1995) - one and a half working generations. To stand out in a world of rapid, relentless and unsettling change is a constant challenge.

Today, the ultimate developmental question for an organisation to ask itself is no longer how do we innovate? But, why do we innovate? This encourages an organisation to answer two further questions: What does it want? How will it know when it has got it? These two questions raise issues related to the consciousness of an organisation e.g., its awareness and understanding of its unique purpose (or reason to exist), and the importance of co-creating a shared vision of its desired future. The problem is either organisations do not know how to discover their uniqueness, or they are unable to creatively harness their uniqueness. This in turn addresses the creativity of an organisation - its ability to actualise its consciousness. This is ever more tricky as we have reached a point in history where traditional and modern sources of meaning are eroding and where the consciousness and creativity of an organisation are inhibited by an array of socio-cultural assumptions (Harman, 1997). These include: materialism (which denies the reality of a living universe), reductionism (which fosters experiences of fragmentation and disassociation), positivism (which shrinks our experience of the world to our five senses), objectivism (which distances us from subjective experience and from building relationships with others, the world and with ourselves), hyper-individualism (which creates consumer cultures and undermines caring, connectedness and community), and economic values (where everything is viewed in monetary terms and discounts the future of any experiences which cannot be quantified and commodified). These assumptions affect how we work and play. We are therefore faced with two socio-cultural challenges: to hospice the death of ways of working which no longer work; and, to midwife the birth of more creative, nourishing, authentic, open and vital futures (Hurley, 1997).

To this effect, we are already witnessing a social and cultural transition with the emergence of a new Integral Culture, a new constructive synthesis of Traditionalism and Modernism (Ray, 1996). The Cultural Creatives who are the bearers of this Integral Culture, by their very nature operate at the cutting edge of change, and share an intimate web of humanistic values. This includes a desire to approach life as a creative work. Similarly, organisations need to learn how to infuse the workplace with more creativity, passion and playfulness. Not to be personally nourished by work, which takes up 60% of our working lives, is no longer a realistic payoff (Leider, 1998).

Integral Culture is inspired, encouraged and supported by the advances in the New Sciences of quantum physics, holistic biology, and complexity and chaos theory, with their discoveries of non-locality, ecological interdependence, and self-organising systems. Our understanding of the world in which we live and work is radically transforming. New science and new technology is dramatically changing society.

To address this new and growing community of inner directeds, (Mitchell, 1997) business is also having to transform, such that the new business of business is to play a creative role in developing vital cultures and contributing to societal learning. The responsibility is inescapable, for business touches everyone. Successful businesses of the future will learn to not only anticipate, but champion, these changes. This requires more than just incremental change in how organisations presently think learn and create, but whole system step-change.

QUANTUM INNOVATION

The word quantum describes a discrete packet of energy. In the metaphoric sense in which the term is used, I refer to the potential of changing (creative) energy into form. In this sense, NPI is a developmental opportunity to align and harness the creative potential of not only an organisation's products and services but also its people. This superpositional (both/and rather than either/or) ambition requires a creative leap, from NPI to Quantum Innovation (QI). This new form steps out of a system of work which no longer works, to one which values the creative spirit, purpose and vision of an organisation. QI plays a vital role in aligning the I-ness of every internal and external customer (the particle aspect), and harnessing the we-ness of an organisation as an organic system (the wave aspect). Unlike innovation for the sake of innovation which usually establishes a precedent for internal conflict and confrontation and external confusion and self-interest, QI is generated through a systemic approach which gets to the heart of an organisation's uniqueness. The consciousness, creativity and design dimensions of an organisation are thereby intimately interconnected in a strategic and vital manner (see figure 1).

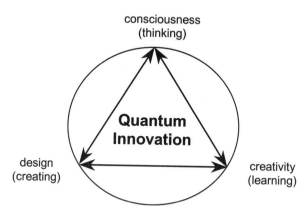

Figure 1 Systemic design development

An organisation, as an organic system, shares similar phenomenological and developmental characteristics with our own individual consciousness and creativity. For example, our own subjective experience has a raw-*ness* or qualia. Qualia are qualities of the mind which are elementary in their nature e.g., the painful-ness of pain, the blue-ness of blue, or the heavi-ness quale of our own physical presence in the world. It has been suggested that qualia are not things but processes, which resonate with our own intimate experience of a stream of conscious-ness. Similarly, discovering and harnessing the unique-ness of an organisation creates a clear, forward sense of flowing time and meaning.

In turn, creativity emerges from consciousness - such that the essence of a creative individual/community is its ability to get to know itself (Meyerson, 1997) - harnessing the creative tension between current reality and a desired reality. Mastery of this creative tension leads to a fundamental shift in our whole position towards reality, where current reality becomes an ally and not an enemy. By unfolding the responsibility of being here in the world, an organisation is able to establish what it knows from what it does not know. Learning and development occurs along this boundary between known and unknown (see figure 2). The higher and more generative the learning, the greater an organisation's capacity to take meaningful, rewarding and strategic leaps out of their comfort zone and into the unknown (Benson, 1991).

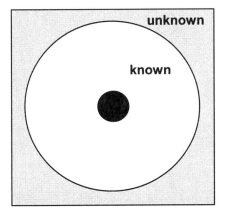

Figure 2 Organisational comfort zone

Rather than rely on words and illustrations, QI is an opportunity to animate a shared vision. Whilst the new business of business is to nourish life, the new and emergent business of design is to mediate this nourishment. By creating interventions (whether product or service based) that maximise learning, encourage vitality and promote a clarity of purpose, QI offers glimpses of living and experiencing the future. QI is therefore an ideal vehicle for replenishing and revitalising an organisation's vision, and for an organisation to replenish and revitalise a society. NPI naturally emerges as an organisation closes the dialogic (free flow of meaning, Bohm, 1987) gap between the here and now, and possible futures. The disciplines of Organisational Development and Organisational Transformation support and facilitate the human dimension of this type of development where there is a genuine commitment to relationships (e.g., trust & leadership), integrity, continuous renewal, wholeness and open-systems.

ALIGNING AND HARNESSING THE FLOW OF MEANING IN AN ORGANISATION

Open-systems theory suggest that all things somehow, someway, link up and influence each other in all directions (Bertalanffy, 1950). This in true in the micro sense when we observe the quantum nature of subatomic wave/particles, and in the macro sense when we discovered the self-organising nature of the universe (Laszlo, 1996).

Like a quantum wave, whereby a flow of meaning resonates through space-time, an organisational vision is like a field of resonating energy aligning the experiences and actions of the workforce towards a meaningful vision of the future (see figure 3). In a field filled space there are no unimportant players (Wheatley, 1992). Fields, although invisible, bring with them the wondrous capacity to turn creative potential into creative form.

The key to renewing creative energy beyond its initial inertia, is an organisation's ability to feel outside the system - to be self-organising, living and open. This is in contrast to a closed system which is unable to interchange energy with its surroundings. Whatever energy it started with will ultimately be used up. Closed organisations move from order and contentment, to chaos and denial. Open organisations move from chaos and confusion, to order and renewal (Janssen, 1982). Open-systems therefore expand indefinitely by importing creative energy and exporting entropy (the inverse measure of the systems capacity to change).

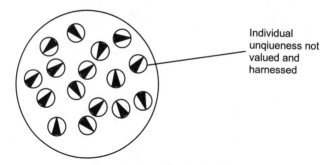

Figure 3a An unaligned, closed organisation

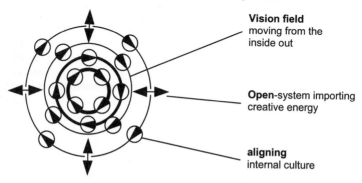

Figure 3b An aligned, open organisation

Whilst a shared vision aligns and inspires an internal culture, and an open system maximising learning, there may still be noise or interference in the space between individuals, functions and levels. In order to develop and harness a frictionless flow of meaning in and around the system, we can refer to the quantum reality of superconductors, superfluids and laser beams which exhibit the highest degree of group agency in the microphysical world (Zohar, 1994).

Unlike glass which has no order and no unity, crystal which has order but no unity, or gas which has unity but no order, superconductors, superfluids and laser beams are highly ordered and highly unified. Each indeterminate particle fills all the space all the time, such that the I-ness of every individual and the we-ness of the organisation can overlap and merge to create a supercoherent entity (see figure 4). Harnessing the creative flow of meaning in and around an organisation helps people to maintain ownership of process and product. QI therefore creates and develops an organisational community which is truly participative, plural, flexible, emergent and purposeful.

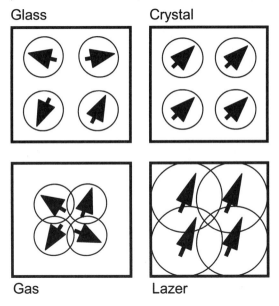

Figure 4 Supercohererence

ALIGNING AND HARNESSING THE FLOW OF MEANING THROUGH AN ORGANISATION

QI works from the inside out as well as the outside in, where everything is relation-al. Indeed, we can only become something more than ourselves through a relationship with others or through experience (a relationship mediated through an object - something other than ourselves and not someone else - with ourselves). It is through such liaisons that we grow and realise our potentialities.

The new business of design is therefore to:
- explore the flux of experience that makes us alive, animated and unique.
- create opportunities for an organisation to share and refine their unique contribution.
- and, in turn enable society to nourish and value meaningful work.

Whilst an organisation's unique purpose enables it to work with the flow of life, a shared vision is the cornerstone to a fully participative workforce, where trust, co-operation and ownership are fully integrated and valued. QI increases the probability of NPI which is not only aligned with the organisations strategic agenda, but which is also autopoietic in its nature i.e., upholding and nurturing its creative integrity. QI therefore stimulates the continual exchange and creation, rather than accumulation of, information, knowledge and experience. The quality and quantity of this exchange is proportionate to the organisation's level of consciousness, or capacity to self-organise. QI therefore generates an emotionally charged design field which flows in and through the organisation (see figure 5).

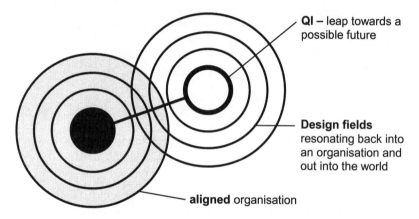

Figure 5 Reflexive waves of possibilities

This reflexive wave of vibrating possibilities fuels the rhythm of growth and shifts the conventional approach to innovation from:
- inventing a corporate mission, to *discovering* an organisation's *purpose*;
- an internal culture which is told to innovate, to one which is *inspired to innovate*;
- an imposed vision, to a *shared vision*;
- conflict and detachment, to *trust, co-operation* and *ownership*;
- limited learning, to *maximised learning*;
- linear thinking, to *systemic thinking*;
- NPI, to *QI*;
- and, a reactive present, to a *creative* and *vital future*.

The new business of design therefore seeks the continual evolution and expansion of potential. By learning to be in genuine relationship with the world and with itself, an organisation can evoke the human spirit and realise its true creative potential.

DESIGN TRANSFORMATION GROUP

Designers of the future will learn to understand wholes. (NB. The word whole comes from the same roots as the world health.) To learn more about wholes and transform our understanding and application of design, the Design Transformation Group (DTG) supports the generation and promotion of meaningful and mindful innovation.

DTG operates as an international learning community - a community of memory, experience and action. DTG shares and co-creates rather than tells and sells; and more importantly, DTG recognises that nothing happens without inner transformation. For example, the impetus for problem solving lies outside ourselves with the intention to make something go away i.e. the problem. Meanwhile, the impetus for bringing something new into existence comes from within. DTG explores the capacity for designers to see and work with a systemic understanding of design - to move from change-fragile to change-agile.

The most important business of new design must be to change our minds.

REFERENCES

Benson, R., 1991, Learning Leadership and the Learning Organisation. In *Visioner & Resultat* (Sweden: SCF Svenska Civilkonomoforengingen), 1, 12-18 & 2, 23-27.

Bohm, D., 1987, *Science, Order, and Creativity*. (London: Routledge).

Harman, W. 1997, *The New Business of Business*. (San Francisco: Berrett-Koehler).

Hurley, T., Healing the social body: pathfinding for the 21st century. Presented at *Beyond the Brain II*, at St. John's College Cambridge, England, August 21-24, 1997, (audio tape).

Janssen, C., 1992 *Personlig Dialektik*. (Stockholm: Liber).

Kim, D.H., 1995 Managerial practice fields: infrastructures of a learning organisation. In, Chawla, C. & Renesch, J. eds., *Learning Organisations*. (Oregon: Productivity Press).

Laszlo, E. 1996, *The whispering pond: a personal guide to the emerging vision of science*. (Dorset: Element books).

Leider, R., 1998, Are you deciding on purpose? In *Fast company*, February/March, 2, pp. 116-117.

Meyerson, M., 1997, Everything I thought about leadership is wrong. In *Fast company greatest hits*, 1, pp. 4-11.

Mitchell, A., 1998, Power to Inner Directeds. In *Marketing Week*, January 22, pp. 24-25.

Ray, P.H., 1996, The Rise of Integral Culture. In *Noetic sciences review*, 37, pp. 4-15.

Senge, P.M., 1992, *The Fifth Discipline: the art and practice of the learning organisation*. (London: Century Business).

von Bertalanffy, L., 1950, The theory of open-systems in physics and biology. In *Science*, 3, pp. 23-29.

Wheatley, M.J., 1992 *Leadership and the New Science: learning about organisation from an orderly universe*. (San Francisco: Berrett-Koehler).

Zohar, D., 1994, *The Quantum Society: mind, physics, and a new social vision*. (New York: Quill).

Multimedia Network Applications in the Fashion Industry

Andrée Woodcock and Stephen A.R. Scrivener

The FashionNet Temin Project (B3004) developed multimedia network applications (MNA's) for the Fashion Industry and evaluated their use over the European Asynchronous Transfer Model (ATM) pilot network. Three applications formed the MNA Toolkit (FINS, Scribble and ShowMe). These enabled pairs of fashion designers to work collaboratively and remotely on the design of new fashion garments. In this paper we describe the project and present results from the usability studies and workshops. We conclude by considering the implications of networked applications for the European fashion industry.

INTRODUCTION

Powerful, computer-supported design applications exist which enable users to perform effectively and efficiently on their own. Much is known about the usability and interface requirements of such systems and in recent years, the fashion industry has begun to exploit this technology to great effect. At the same time the fashion industry has become increasingly internationalised such that garments are often the product of combined input from many different countries. For example, a garment might be designed in Britain of fabric sourced from India, printed in Japan, and manufactured in Spain for a client to retail in England. Design, manufacturing and marketing of garments involves a high degree of organisation and control if clients and providers in the fashion pipe-line are to maintain quality and competitiveness. Computer applications and networks are now being developed which will revitalise the industry by allowing fast, effective and efficient communication between SME's.

Fashion is a highly visual enterprise. Fashion garments are physical artefacts with shape, colour, fabric, structure etc. and many decisions have to be made in the presence of these objects or representations of them. The industry will need to be supported in the future by integrated, broadband (high capacity) communication networks supporting teleconferencing, image transfer and manipulation. At present there is little understanding of the requirements of the fashion industry or the technological infrastructure necessary to satisfy these needs because, to date, only narrow band communication networks have been available.

THE AIMS OF THE PROJECT

The FashionNet consortium (see Table 1) comprised 12 telecommunications providers, applications developers, research and user organisations from Ireland, Britain, Portugal and Germany, funded under the European Commission's TransEuropean Network - Integrated Broadband Communication (TEN-IBC) programme. Over a three year period the team developed new methods for designing, implementing, testing, and evaluating multimedia network applications over ATM - both point to point and multipoint under different bandwidth configuration (1,2,4 and 8Mbps). In addition, ways of assessing the impact of the technology on job satisfaction, business costs and organisational structure were developed.

The overall objectives of the project were to assess the impact, usefulness and usability of these types of MNA's for the European fashion industry, and feed results back to system developers, network providers and members of the fashion industry. These were fulfilled by a series of local and international trials in which fashion designers were invited into computer laboratories to use the applications in realistic scenarios running over different configurations. The usability of the applications was measured using standard Human Computer Interaction (HCI) tools, and the effects of the bandwidth configurations were evaluated using breakdown analysis. The implications of the way of working for the fashion industry were evaluated through cost benefit assessment (Eason, 1992).

THE MULTIMEDIA NETWORK APPLICATION (MNA) TOOL-KIT

The MNA tool-kit consisted of three separate applications for fashion industry users (such as designers, managers, clients, buyers, and marketers etc.). Using these applications up-to-date fashion information could be accessed (such as fabrics, prints, fashion shows and manufacturers' products from FINS) and remotely located workers could design, modify, present and discuss designs using a collaborative GroupWare. The applications are described below.

Scribble[1]

Scribble is a collaborative annotation tool which allows up to six users, located on physically separated workstations, to draw and annotate text over an underlying prototype garment design, a fashion sketch, knit, print design or a digitised photographic style image. Scribble supports the input of high resolution pictures, annotation and sketching. Typical drawing tools are freehand and line drawing, circles, ellipses and rectangles, both filled and unfilled. Graphics created by Scribble can be saved to disk for further post-processing by third party packages.

---

[1] Scribble was developed by ADETTI, Portugal and Trinity College Dublin.

Table 1 The Fashion Net Consortium

PARTNER	EXPERTISE	COUNTRY	ROLE IN PROJECT
ADETTI	Telecommunications/soft ware developers	Portugal	Telecommunications, Application (Scribble), End user trials
Franhofer – IAO	Applied Research	Germany	Evaluation, End user trials
Queens University Belfast	Academic	N. Ireland	Telecommunications, End user trials
Trinity College Dublin	Academic	Ireland	Telecommunications, Application (Scribble) End User trials
Ulster University	Academic	N. Ireland	Evaluation
Derby University	Academic	UK	Evaluation
Nottingham Trent University	Academic	N. Ireland	Application (FINS)
Portuguese Telecom	Telecommunications	Portugal	Telecommunications, Management
Al Ferano	Fashion Industry	Germany	End user company
Maconde	Fashion Industry	Portugal	End user company
Ski and Sports	Fashion Industry	N. Ireland	End user company
Coats Viyella	Fashion Industry	UK	End user company

FINS[2] (Fashion Intelligence Navigation System)

FINS is a multimedia database for the clothing and textile industries providing information on designs, trends, suppliers and business skills. Its users have access to a highly accurate and comprehensive body of product and market information, for reference. It has a rich multimedia environment with high resolution pictures, presentations and video clips complementing textual information reflecting the highly visual nature of the fashion industry. As the information it contains is time-sensitive the database is updated regularly.

Video conference - ShowMe[3]

ShowMe comprises a set of network conferencing tools for users at different sites. It supports video and audio teleconferencing, whiteboard and application sharing. FashionNet was one of the first projects to use ShowMe over a wide area transnational ATM network. The video quality offered is dependent on the network bit rate available and also upon the colour depth of the video display. The video capture parameters may be altered to modify the demands on network and hardware performance - e.g., using more / fewer colours, using a monochrome scheme, sending different raster resolutions and varying the maximum frame rate.

---

[2] FINS was developed by Nottingham Trent University. Now available as a service on the Internet
[3] Developed by SunMicrosystems

NETWORK CONFIGURATION

The Toolkit was tested over a local ATM at Queen's University Belfast (Woodcock and Scrivener, 1997a) before being used in a series of international trials connecting users in four European sites - Belfast (QUB), Dublin (TCD), Lisbon (ADETTI) and Stuttgart (IAO), see Figure 1. Using this network it was possible for example, for a designer in Lisbon to work with a colleague located in Belfast. 12 international trials were conducted using different configurations of the system, and with designers from the four different countries.

THE DESIGN SCENARIOS

Three different, realistic scenarios were used in the trials. These were chosen to demonstrate the potential for co-operative working using the MNA toolkit and allowed users to work on realistic, and interesting tasks which made use of all the applications. All trials were video - recorded for later analysis, and the participants were required to complete questionnaires on the completion of the trials. The three scenarios were:

Sportswear for cyclists. The national cycle team would like to increase their 'visibility' in future cycling championships such as the Tour de France and the Milk Race. Their integrated sportswear should take advantage of new fabrics and the design should reflect the idea that cycling promotes health and fitness. In the design you need to specify the garment manufacturers, the type of material the items will be made out of, and the badges of the sponsors.

Women's top. At a recent fashion show you were very much impressed by the stripy women's top designed by Clements Ribeiro. Using this as a starting point, design a similar top in a liquids palette, which might be used for leisure wear. Bear in mind that the fabric should withstand a lot of washes without losing its shape or colour.

Boys wear. Design a co-ordinated winter costume (trousers and top) which will appeal to 8 - 11 year old boys. It should reflect the latest trends, be durable and allow them to pursue their normal outdoors activities (e.g., football, running, cycling).

The sessions lasted for two hours and required remote pairs of designers (or in some cases two designers and a 'client') to work co-operatively to produce an on-line portfolio, meeting the requirements of the brief. Subsequent analysis focused on the ability of the applications to support remote working when run over different bandwidth configurations. A review of the methodologies employed in the project is outlined below, although in this paper we will only present results relating to user cost benefit assessment and system evaluation.

THE EVALUATION METHODOLOGY

During the trials the bandwidth available to the applications was varied. In the local trials 1,2,4 and 8 Mbps were used, however for the international trials, only three higher bandwidth conditions were used, as 1Mbps was found to be too low to support co-operative working of this nature.

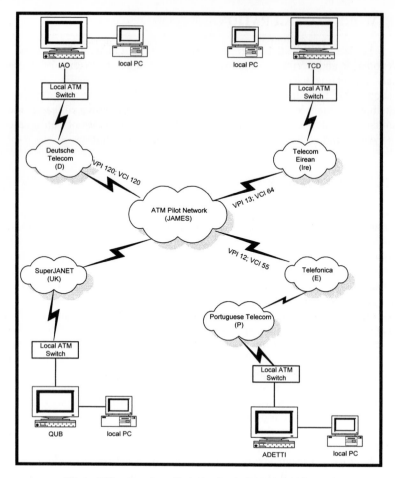

Figure 1 The network configuration for the European trials

All other variables were kept as constant as possible, for example, by producing packs outlining experimental procedures to be undertaken at each of the four computer laboratories, providing participants with training on the applications and only using practising fashion designers. The evaluation techniques (outlined in Table 2) provided a complete picture of the use of the MNA toolkit from a technical and a user oriented perspective.

Table 2 Summary of evaluation techniques

TECHNIQUE	PURPOSE
User cost benefit assessment (Eason, 1988).	Investigate issues affecting acceptance of the technology
HCI evaluation (Ravden and Johnson, 1989).	Establish baseline measures of application performance, and also to study the effects of varying the bandwidth on the performance of the applications.
Attitudinal scores (Badagliacco, 1990).	To determine whether the performance of the applications effected the users attitudes towards computers
Breakdown analysis (Urquijo et al , 1993).	A measure of perceived performance of the applications and indication of where the problems occur
Bandwidth analysis recorded using OpenView/ ForeView	This provided information on the amount of available bandwidth used by the applications

RESULTS

Technical results have been reported in Marshall and Woodcock (1997). In this section we present the results from the usability analysis and the cost benefits assessment, as these are believed to be of central importance to new product developments in the fashion industry.

Computer – supported cooperative work (CSCW) and fashion design

Attitudes towards MNA's and computer support for fashion design were gathered using cost benefit questionnaires, and interviews with the representatives of the fashion industry who attended the international workshops. These showed that fashion designers and managers were very keen to use CSCW and could see the benefits of using the technology to enhance the design process.

The FINS database was seen as an important innovation, which would reduce the amount of time and expense required to source information in the early stages of the design. Information and images from the fashion shows and on style and colour trends can be used to shape the initial concept, and the information on manufacturers and suppliers, include contacts, manufacturing capacities and ranges produced, can help in speeding up the production of the final market. For small design companies (such as craft and cottage industries) having their details included in the database will greatly enhance their corporate profile and provide an in-road into electronic commerce.

Scribble was useful as a means of communicating ideas and concepts, providing an opportunity to design on-line and collaboratively, using functions not normally available to designers. However, some concern was expressed as to whether designers did work together during the early stages of the design. In terms of designer–to–client communication, Scribble could reduce the design lifecycle and help remove misunderstandings by providing a means of conveying and jointly editing the concept design. In achieving this communication, ShowMe was seen as equally important in enabling discussion of the design brief and to resolving misunderstandings early in the process.

Although the views expressed were overwhelmingly favourable concerning the applications and the MNA Toolset as a whole, reservations were expressed concerning whether designers would like or could use the technology. The adoption of CAD systems by the fashion industry has to be considered in relation to retraining, financial rewards and possible redundancies. These issues, along with financial factors, network availability, interoperability and adoption of CSCW by a whole supply chain will shape the rate at which the technology is integrated into the European fashion industry.

Usability of the applications

The usability of each application was assessed when it ran independently and as part of the MNA toolkit in the context of the international trials. It was hypothesized that any usability problems found when the applications were run under optimum conditions were likely to be exacerbated over reduced bandwidths.

With regard to the shared drawing and annotation application, Scribble the main usability issues related to incomplete functionality, speed, difficulty in commands and lack of user help. Running under reduced bandwidths (especially 1 and 2 Mbps), Scribble was not robust enough to guarantee fast image transfer and synchronous activities were impossible. A technical solution to this problem was found at 4 Mbsec, by allowing a greater proportion of the bandwidth to be made available to the application (but at the expense of the ShowMe).

Usability issues relating to FINS related mainly to the lack of user help and the underpopulation of the database (which was still being expanded at the time of the trials). This meant that users were unable to establish a clear understanding of the underlying structure of the database and had navigational problems. This finding was corroborated by the international trials where designers found it hard to develop a mental model of the database, and co-ordinate their movements. Image transfer from FINS into Scribble proved slow and unreliable.

The usability of ShowMe could only be assessed during the international trials. When the system was downgraded to allow Scribble to work under reduced bandwidths, users did not notice reductions in image quality (e.g., number of colours, frame rate), as the video channel was only used for confirmation of the presence of the partner. Degrading the sound quality, on the other hand, had severe effects on the designers ability to continue with the trial.

Whilst the questionnaires provided an overview of the usability problems they did not capture the effects of these on design activity. Breakdown analysis provides diagnostic information from rich data (such as video and audio recordings). It is a systematic and fast means of approaching large quantities of communication data, identifying those areas which highlight problems, and relieving the evaluator of the task of consulting or becoming an expert in a more complex form of conversational analysis or HCI. A breakdown can be defined as 'the moment when the user becomes conscious of the properties of the system and has to mentally breakdown or decompose his or her understanding of the system in order to rationalise the problem experienced' (Scrivener *et al.,* 1996).

It has been customary to classify breakdowns in terms of the TUTE framework, Task, User, Tool, Environment, (Shackel; 1981, 1991) with the aim of enhancing the quality of the information regarding its nature. The interactions which result from the employment of this model can be summarised as follows (see Table 3).

The Toolkit can be interpreted as a system enabling fashion designers to work together in the production of the artefacts, namely an online portfolio (of one or more designs). User breakdowns are events in which the capacity to produce, transmit and receive 'design products' is inhibited. These breakdowns can be attributed to the applications, the effects of the bandwidth, or on user -user problems. By analysing these events users and software/network designers can better understand the implications of poor usability and functionality.

A Delphi event recorder was developed to aid in the recording, timing and classification of events. All occasions in which a user expressed or appeared to have a difficulty were recorded. Verbalisations occurred naturally within the design session and some of these related directly to usability issues (e.g., poor pen design). At other times breakdowns had to be inferred from user behaviour (e.g., repetition to compensate for poor audio quality). Each breakdown was categorised in terms of:

- Its relative importance in the context of the trial where 'high' referred to a major breakdown interfering with the way the task was performed, 'medium' was a cause of user confusion, requiring adaptive behaviour but not causing its abandonment, and 'low' was an isolate incident of short duration which did not significantly interfere with the task.
- Attribution of the event in terms of the TUTE model.
- The long and short-term effects of the problem on design activity. The effects of the breakdowns were manifest in standard HCI usability problems (e.g., insufficient functionality causing frustration and changes in working practices), and in inhibiting their ability to work together (in severe cases resulting in abandonment of the session). If we take the case of bad echo (caused by latency) - a short term effect might be that the user shortens the sentence, rate of speech production is slowed, as the user waits for the echo to stop before embarking on the next phrase. A long term effect might be to inhibit utterances, stop discussion and spontaneity. Clearly system design should aim to minimise detrimental long term effects.
- Solutions undertaken by users. There was some consistency between strategies employed to overcome problems e.g. breakdown of audio channel resulted in three different solution depending on severity; change speech production (slower and louder); change communication channel (used written rather than verbal communication); abandon trial.

Table 3 Model used to describe breakdowns

CATEGORY	GROUP	EXAMPLES
User and Task	Difficulties in understanding the task	Comprehension
	Difficulties in accomplishing the task	knowledge, skill to perform the task
User and Tool	Hardware	1.tool failure e.g. system crash
		2. understanding or unsuitability of tool e.g. pen not touch sensitive
	Software	1. tool failure e.g. software hang up
		2. understanding or unsuitability of tool e.g. no fill function on a drawing package
User and Environment	user becomes conscious of the properties of the environment	talking in the room where the user is working
User and User	Sufficiency	information received by the partner is not sufficient for understanding the sender's intention
	Clarity	message is inaudible or incomprehensible
	Comprehension	cultural differences may lead to misunderstandings
	Attention	receiver is absorbed in the task
	co-ordination	users fail to co-ordinate their utterances and interrupt each other
	Feedback	message sender does not receive any information from the recipient.

Breakdown analysis provided a richer and more accurate recording of usability problems than the questionnaires (Woodcock and Scrivener, 1997b). It clearly identified the nature of the problem, the impact on the overall quality of the interaction and the ability of the designers to perform the task using the tools at their disposal. The use of breakdown analysis in the manner suggested i.e. using it to look at adaptive behaviour on the part of users demonstrates those cases when there is a need for software intervention. These cases would be those in which a problem, or a solution to a problem leads users to behave in an 'unnatural manner' (e.g., shouting) which increases cognitive load, reduces interaction (because the quality of the sound is so bad), and impedes the design activity.

In terms of the FashionNet Toolkit, breakdown analysis confirmed that for FINS the main problem was in understanding the structure of the database, finding the required information, and knowing that the only solution had been found. Although the interface provided information on the number of pages matching their query this was disregarded. Users searched through the same pages repeatedly, and never understood the design of the database. Navigational aids such as bookmaking and a notepad might have helped. If FINS is to be used for sourcing information, the system developers need information on how it is used during a real task.

Scribble was not a robust application. File transfer was slow and unreliable. It was useful as an annotation system (e.g., for verification and adaptation of concepts) but did not support joint drawing, especially over lower bandwidths. The severest problems faced by users was refreshing of screens (up to 15 times in succession) which did not relate to any actions being undertaken by the users and the lack of synchronicity.

For the video conferencing system the major finding was that audio quality should not be compromised. Although previous results have shown the importance of supporting non verbal cues (such as expression or eye movement) in our project this was not important. The video channel was used merely for confirmation of the presence of the other designer.

DISCUSSION

The FashionNet project sought to involve users in the early design and testing of applications and networks in the context of real scenarios. In many cases the technical difficulties (such as interoperability issues and network issues) caused usability problems to be raised which are normally transparent to the user. For example successful completion of the trial required significant involvement of technical staff, in reconfiguring systems, rebooting machines, and setting up network connections. However, the fashion designers were adaptive and tolerant in their use of the applications, and were able to look beyond the present instantiations of the applications to see the wider implications of computer supported co-operative design. The severity and frequency of the breakdowns indicated that although design work can be conducted under highly restricted bandwidths, it is not recommended.

Of the methodologies employed, breakdown analysis proved the richest, most useful and reliable information for system development. It should be remembered that although a lot of studies have been conducted on the use of co-operative working with text based systems, comparatively few have been conducted which deal with multimedia systems. Software developers do not know the functionality which should be included or how images are used to communicate and develop ideas. The earlier this can be incorporated into software development the better. The applications provided in the MNA Toolkit, are going to be essential to fashion companies seeking to engage in electronic commerce in a global market. Fashion designers will need to communicate and discuss concepts with clients quickly and over a distance if they are to shorten the design life-cycle and increase market penetration. This will require a system like Scribble, supporting joint annotation of sketches and a video conferencing tool allowing communication between all members of the design and supply chain. Effective, efficient and fast sourcing of information is vital to capture new markets and predict future trends. In the future, virtual fashion shows will become more commonplace. The FINS database provided designers with an insight into the way in which material might be sourced, and suppliers and manufacturers located.

CSCW will revitalise the fashion industry, but realising its full potential will be a gradual process. Firstly, it requires the development of integrated systems to support fashion design process, based on an understanding of designers requirements. Secondly, it requires commitment from individuals and organisations to retrain (as appropriate) and use the systems on a daily basis. Lastly, it requires a robust and cheap network to enable all members of the fashion pipe-line (customer, designer, supplier, manufacturer and distributor) to work together in a co-ordinated manner.

REFERENCES

Badagliacco, J.M., 1990, Gender and race differences in computing attitudes and experience, In *Social Science Computer Review*, **8**, (1), pp. 42- 63

Eason, K. D., 1988, *Information Technology and Organisational Change*, (London: Taylor and Francis).

Ravden, S. and Johnson, G., 1989, *Evaluating usability of human-computer interfaces*, (London: Ellis Horwood).

Scrivener, S.A.R., Urquijo, S.P. and Palmen, H.,1996, The Use of Breakdown Analysis in Synchronous CSCW System Design. In P. Thomas (Ed.), *CSCW Requirements and Evaluation*;(Springer, London).

Shackel, B. 1981, The concept of usability: in J.L.Bennett, Case, D., Sandelin, J. and Smith, M. (Eds.) *Visual Display Terminals: Usability Issues and Health Concerns*, (New Jersey: Englewood Cliffs, Prentice-Hall), pp. 45-88.

Shackel, B. 1991, Usability-context, Framework, Definition, Design and Evaluation. In Shackel, B. and Richardson S. (Eds.), *Human Factors for Informatics Usability*, (Cambridge: University Press, Cambridge).

Urquijo, S.P., Scrivener, S.A.R. and Palmen, H., 1993, The use of Breakdown Analysis in Synchronous CSCW System Design. In *Proceedings. of the Third European Conference on Computer-Supported Cooperated Work*, Milan, September, pp. 281-293.

Woodcock, A. and Marshall, A., 1997, *Evaluation report on the Trial: Deliverable, D6: B3004 Final Project report,* submitted to the EU in January 1997, available from the authors.

Woodcock, A. and Scrivener, S.A.R., 1997a, User Requirements of Networked Applications.In Robertson, S., (Ed.), *Contemporary Ergonomics*, pp. 516-521.

Woodcock, A. and Scrivener, S.A.R., 1997b, Evaluation of User Requirements for Broadband Communication in the Fashion Industry, *13th Triennial Conference of the International Ergonomics Society, Tampere,* Finland, June 1997, Vol **7**, pp. 587-590.

Putting Practice into Practice: assimilating design into the corporate culture

Martin Woolley

This paper explores the nature of organisational change that resulted from establishing in-house industrial design expertise within a company manufacturing capital equipment. The mechanism that brought about the change was three-year funding provided through an alpha-rated Teaching Company Scheme (TCS), in which the author was the academic supervisor.

Following brief details of the UK Teaching Company Scheme, the way the project aims related to the strategic thinking of the company as a whole is then outlined. Reference is made throughout of the approach of the project when engaging with management criteria, product planning, marketing, research and development.

Several design related factors are identified, which acted as the primary drivers of change: including the requirement to diversify the product range due to the disappearance of traditional markets, the transferability of existing product technologies and the identification of new marketing opportunities. Specific reference is made to the design role of IT both as a facilitator of corporate change as well as a design modelling, visualisation and decision making tool. In addition, introducing in-house design expertise brought about a questioning of the keystone to change, the nature of the management commitment to design. The positive elements are contrasted with the factors that acted as barriers to change, for example: the traditional perceptions of company objectives, the external economic climate, and internally - management communications and design credibility.

Evaluating the results of the project with reference to short term design outcomes and long term organisational impact, there is the question of the role of industrial design in relation to a highly regulated capital equipment sector, with its own design agenda (which is different from the more frequently addressed design of consumer goods). In this context the redefinition of the 'user' in relation to capital equipment is discussed, along with an appraisal of the aesthetic and functional criteria relevant to the capital equipment sector. These issues are examined linking the approaches and outcomes of a specific case study to a wider review of contemporary design management thinking, and the implications for design education to be addressed for future projects.

The principle argument advanced is that the successful adaptation and exploitation of design as a resource is reliant upon a combination of attitudinal, cultural and structural adjustments within a company, this operates in parallel with a broadening of the traditional industrial design role of the company by the incoming design practitioners.

In conclusion, a single case study may only contribute to a generic hypothesis, however the issues raised may be relevant to contexts in which design practice engages with highly functional criteria that are also associated with emerging technologies. The paper is aimed at those participating in or managing such processes - whether as managers, design managers, design practitioners or members of multidisciplinary teams with a design component.

INTRODUCTION

"Whilst competitive advantage can come from size, or possession of assets, etc. the pattern is increasingly coming to favour those organisations which can mobilise knowledge and technological skills and experience to create new products, processes and services" (Kay, 1993).

Concern about the UK's ability to translate innovation, invention and discovery into commercially successful goods and services has been identified as a key component in a considerably reduced UK manufacturing base. To stem this decline (in wealth generation and the country's economic viability) has been a long-established fixation of successive governments. As a result, many different governmental strategies for improving performance have been implemented, with varying levels of success. Increasing emphasis has been placed on the encouragement and funding of projects that bring together academic and commercial partners. The field of design, particularly industrial design has begun to make a significant contribution to such projects. Part of this progression, the relationship between organisational change and successful adaptation: the ability to assimilate 'design' through teamwork and cross functional process improvement into the corporate culture and corporate strategy, are today viewed as the key elements in improving performance and business survival.

A related theme within academic circles is the relative importance of basic, strategic and applied research - a debate pitched between the opposing poles of 'discovery for its own sake' versus rapid commercial return on research funding (Woolley, 1998). In practice a range of targeted government funding mechanisms now exist along this spectrum. One well established source of funding for applied research is the Teaching Company Scheme (TCS) which supports higher education institutions to work in tandem with the developmental needs of commercial companies. The arrangement encourages a two-way transfer with mutual adaptation, so that academic institutions and businesses work together, each informing the others practice. The primary aim is economic - 'ivory towers' applying research and analysis to relevant real-world problems, and the business sector using and adapting well-researched academic models to become fitter and more able competitors in the global marketplace.

THE UK TCS PROGRAMME

Since 1975 ten UK government bodies have sponsored TCS, which supports partnerships between companies and universities for technology transfer and training. TCS's aim to:

- Enable companies to make strategic business advances through projects which would otherwise be beyond their resources of knowledge and skilled manpower.
- Assist the development of able young graduates into future industrial leaders by providing high profile training in industry with academic support.
- Enable universities to develop their staff and improve the quality and relevance of research and teaching by applying their knowledge to exacting industrial projects.

Aims for the companies within TCS partnerships include, introducing:

- New or improved products, services and processes.
- Improved systems.
- New markets or improving penetration of existing ones (Teaching Company Directorate, 1998).

TCS partnerships are financed by government grants made to the university partner, complemented by financial contributions made by the partner firm. The funds support employment of one or more graduates, known as 'Associates', who work in the partner company on projects designed to realise the advances sought by the firm. Associates are supervised jointly by a senior company employee and senior academics from the university. In addition to practical learning, Associates receive formal training aimed at developing communications and managerial skills. Around 600 companies - of all sizes and in a wide range of sectors - participate in TCS. It supports the training of about 1000 graduates and involves almost every UK university.

The formula has proved highly successful with over 2000 schemes completed to date, it centres on project programme agreed by the academic and commercial partners, which is carried out almost entirely within the company, to achieve goals related to the companies' developmental requirements. The associates, generally young graduates with a first, or higher degree, are employed as members of staff of the academic partner institution to carry out the agreed work programme. It is generally agreed that Associates are placed at a senior level within a company structure to provide impetus for design concepts to be developed and progressed. The institution provides one, or more, members of academic staff to supervise and act as a conduit for relevant knowledge, expertise, skill and technology transfer into the company. Although the primary aim is to improve company performance, there are the significant benefits of graduate employment / training, and the opportunity provided for 'on the job' live investigation by academic staff.

TCS programmes are geared to commercially defined ambitions, they require a substantial level of investment, in time and resources and do not represent a 'quick-fix' solution. The duration of a scheme generally ranges from 1 to five years or more, and will involve one or more Associates for two-year periods each. The targets of the agreed programmes are generally weighty and often involve a combination of strategy, management, investigation and innovative development.

The changes brought about by the programmes may affect the approach, strategy and ways of working throughout the whole organisation.

Originally the focus of TCS programmes was exclusively science / engineering / technology-based, but over the years, the agenda has broadened considerably to embrace all areas of operation including business studies, marketing, management and latterly design. A number of design-led TCS programmes have now been completed throughout the UK in a wide variety of manufacturing, servicing and consultancy companies. Currently there are a number of interesting comparative studies of design-led TCS programmes emerging which reflect the impact of design on a range of company defined problems (Design Council, 1998).

This paper focuses on a design-led scheme for which the author was the academic supervisor whilst based at the University of Central England in Birmingham, working with the Poole-based company Zellweger Analytics Ltd. The mechanism for change was a three year funded and ultimately alpha-rated TCS programme. Assimilating design into the corporate culture at Zellweger provided an opportunity to explore the nature of organisational change resulting from establishing in-house industrial design expertise within the capital goods sector.

THE COMPANY

An international company, Zellweger (originally Sieger Ltd.) is one of the world's leading manufacturers of environmental monitoring equipment for combustible and toxic gases. The company develops, manufactures and markets internationally, instruments and instrumentation systems for oil and gas, mining, petro-chemical, producer and user industries. The products range from hand-held instruments to large-scale alarm or 'shut-down' safety systems, involving some hundred sensors in fixed locations coupled by a digital communications network to central control instrumentation. The products are designed to conform to strictly defined safety standards and are certified for safety of operation in hazardous atmospheres by internationally recognised test and certification authorities.

When the TCS scheme was first discussed, the company was facing a strategic challenge related to fundamental changes in its traditional customer-base and increasing international competition. To meet this challenge the need to develop new products and exploit new markets was identified. At a senior level, the company recognised the need for change and believed that amongst several strategies, design might act as both a catalyst for, and facilitator of, such change.

Prior to discussions, the company had employed design consultancies to redesign a small number of existing products and contribute to new product development. Unfortunately, the results of the consultancy projects were regarded critically by a number of senior staff. It was felt that the products were unnecessarily difficult and costly to manufacture, there was a commonly held view that this had resulted from a lack of understanding of the company - its technological base, markets and cost management. As a result, it was generally believed that if a concerted effort was to be made to introduce industrial design, it should be from within, as an in-company design unit, rather than externally - using consultants. This thinking ultimately lead to the view that a TCS programme could act as the mechanism for establishing and developing an in-house industrial design unit. The TCS could provide specialised academic expertise in industrial design together with qualified individuals who would introduce and integrate industrial design over a three-year period.

PROJECT HISTORY

"It was found that 98% of (TCS) programmes with cohesive objectives were in the high scoring group, whereas 89% of programmes with fragmented objectives were in the low scoring group. The analysis also suggested there is another inter-relationship between various inputs, i.e., clear programme objectives, relating closely to a company's strategy which are supported by relevant academic expertise, usually generate company commitment and successful outcomes" (Senker J., *et al.,* 1993).

The partnership resulted from a series of introductions between the author, the Company and the regional TCS consultant. From the late 1980s I had been aware of TCS programmes and that at least two programmes had already centred on design issues. Indeed a TCS programme appeared to be particularly appropriate for design because of the emphasis on applied, company-defined opportunities rather than the projects favoured by the UK research councils which tended to be less 'close to market'. This inclusive view of the schemes' relevance to design was later supported by the conclusions of a Science Policy Research Unit (SPRU) report commissioned to evaluate TCS, which suggested that TCS programmes can be successful for large and small firms. Moreover, the authors have been unable to find any evidence to indicate that any sector of the economy, or any type of academic expertise, should be excluded from consideration as the basis for TCS programmes (Senker, *et al.,* 1993).

The project had a lengthy gestation period, as both the academic and commercial partners were new to TCS; there were initial Associate recruitment difficulties and the proposal required substantial refinement in response to a deeper diagnosis of changing company needs. A key player within the company at the time was a director and scientific consultant who had considerable knowledge of, and enthusiasm for, TCS and had previously acted as a UK TCS adviser. Two other senior members of staff - the Director of R&D and the Manager of the Design Services Department - played a crucial role in early meetings and helped to shape its primary focus — to create an in-house industrial design unit which would substantially contribute to existing and new product development with the appointment of two design Associates. The programme was eventually established for three years, to employ two Associates each on a two-year contract, overlapping in year two. It embraced a range of general design issues from existing product re-design and design for new product development to company review and computer-based design development.

From the beginning, the company had displayed an enlightened view of the professional development required for the Associates, it was made clear that they would be working at a senior level within the Company. The senior management wished to evince change, and it was agreed that to have impact the Associates should understand and respond to policy making in order to facilitate the changes required.

ATTITUDINAL INFLUENCES IN RELATION TO THE TCS ASSOCIATES

At an early stage, the programme revealed key factors in relation to the recruitment and selection of associates. It was common practice to draw up a 'specification' for Associates as part of the original TCS proposal. A dilemma is, that the TCS programme is geared to comparatively inexperienced graduates, often it is their first job, but they are 'fast tracked' to senior positions and often deal with leading-edge, complex and strategic issues. Maturity, diplomatic and management skills are all requirements for stimulating change within this context. Thus recruitment focused on a combination of design and interpersonal skills with the added criterion of IT skills with regard to the second Associate.

TCS recognises that recent graduates may lack organisational experience and communication skills, so an intensive introductory course is provided with these components well represented. The deployment and support of both an in-company and academic supervisors also partially compensates for professional inexperience. When considering the stimulus to cultural change, to accommodate new design resources - this issue highlights the need for dual abilities: able designers who are also equipped with the competencies to assist with engendering change, as illustrated in Table 1.

Table 1 Associate attributes which engender change

The ability to:
• 'open doors' – establish and develop working relationships across the company
• develop overview of company structure, individual functions, markets, technologies & design requirements
• operate as a team member
• develop strategies for diffusing design across the company
• promote design as a valid and useful discipline within the company
• practice design in a wide range of contexts within the company

A relevant factor can be traced back to design education. Over a period most, if not all, design courses recognising that design takes place within a business / organisational context have provided matching course content. However, role-play, critiques and formal teaching can only go so far in developing the personal skills necessary to negotiate a career within this context. The traditional educational emphasis on design as essentially a personal voyage of creativity, skills and knowledge acquisition, tends to develop a commitment and dedication to the design profession, almost as an end in itself. Hence it is possible for graduates to enter their first employment with a stronger commitment to practice than to the company.

It might be argued that given contemporary employment patterns, this is not necessarily undesirable, a sense of vocation is needed to steer a path through not infrequent job changes and fluctuating consultancy work. However, it fails to address the inevitable need for professional adaptation that career development often provokes or demands. The assumption is that progression occurs along discreet career paths with the acquisition of new skills and knowledge along the way whilst retaining a professional identity defined by higher education. This model, does not take into account that practitioners may be called upon, or choose to change their career pathway to such an extent that substantial professional re-definition occurs: footballers sometimes become managers, teachers become administrators, trade unionists become politicians. It could be argued that too great an emphasis upon the lifelong vocational 'mission' of design may ultimately reduce the individuals ability to adapt to the real-world professional context. In terms of the Zellweger scheme it was apparent that Associate success was largely determined by the ability to adapt to the organisation rather than by exercising design competencies in isolation.

DESIGN AND CULTURAL CHANGE

The principle argument advanced is that the successful adaptation and exploitation of design as a resource is reliant upon a combination of attitudinal and structural adjustments within a company, this operates in parallel with a broadening of the traditional industrial design role by the incoming design practitioners.   Thus it necessary to define and explore a problem possibly common to other design-based programmes, namely that cultural changes are required to bring design in-company with the corresponding adjustments required of the incoming Design Associate. Senker supports this view in a wider TCS context and seeks to identify significant factors in programme successes making it clear that in some cases, organisational learning resulting from TCS programmes has been so extensive that it is justifiable to refer to it as 'cultural change' (Senker *et al.,* 1993). In this context, the Zellweger TC programme was intended from the outset to promote cultural change as well as introduce new design competencies into the company. The following explores how, and if, this was achieved.

CONSOLIDATING CULTURAL CHANGE

A recent paper by Svengren (1998) postulates that design management takes place on three levels: "on an operative level, like project management; on a strategic level, e.g. strategic concepts related to identity questions; and on a philosophical level, i.e. how management value design as a resource and basis for everything that is decided upon and done in the company." As the TCS programme in this case sought to implement industrial design as an in-company and therefore 'managed' resource, it was also required to operate on these three levels. It is equally clear that such an in-depth process will not occur instantly but may require introduction by stages. The model embodied within the original TCS proposal contained a number of key stages, which, in retrospect, constitute the primary stages in the phased introduction of managed industrial design into the company. Although introduction of design personnel within a developed company for the first time, requires a process of adaptation which could be protracted; however, the company is more likely to be forced into rapid change by the need for immediate resolution in the light of pressing external economic factors. Such a situation requires a process of significant bridge building from both 'sides'. This process illustrated in Diagram 1 and defined in Table 2 comprises four distinct, overlapping phases.

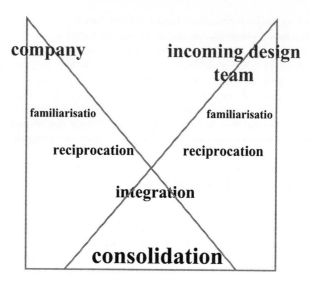

Diagram 1 Developing an industrial design culture - bridge building

Table 2 The design consolidation process

Phase:
1. **Familiarisation** – The initial stage in which the incoming designer(s) first make contact with the company and absorb information about its structure and context. The company increases its awareness of design its relevance to the company.
2. **Reciprocation** – the direct exchange of information between the designer and the company – demonstrating competencies, benefits, comparing views, considering working relationships.
3. **Integration** – design and other product development tasks are combined as a team or group effort and the parties move together within the wider management structure.
4. **Consolidation** – the final stage, where integrated working processes are fine-tuned to meet the strategic, long term aims of the company.

Phase 1: familiarisation

The TCS programme, at the planning stage, recognised the first phase of the programme for both the Associate and the Company as being a process of mutual familiarisation. For example, the first Associate at Zellweger Analytics was required to analyse the design strengths and weaknesses of the company and its competition in the field and the second Associate conducted an IT-related design audit of the company.

In parallel, as indicated in Table 3, there was a need for a familiarisation process within the Company to absorb the purpose and methods of the TC programme and the potential contribution of industrial design. The process was engendered by early meetings with an external management consultant with long-term experience of the company and acted as a mediator between design associates and supervisors. As a result, to raise awareness of design throughout the company, an information provision programme was developed employing the use of newsletters, exhibitions of work and presentations.

Table 3 Mutual familiarisation.
The primary areas of initial information transfer between the design team and the company

Design Team	**Company**
Company history, structure & resources	Design knowledge, skills, methods, experience
Company people (especially engineering & marketing)	Design potential with reference to the company
Competition	The management of design
Markets	Design aims, role and function within company
Production methods	TCS aims, methods and management
Production technologies	

Phase 2: reciprocation

The period of reciprocation marks the implementation of the design process, initially on the basis of one / two-way exchange of services: early meetings to discuss and negotiate ways of working together effectively, briefings to establish action tasks. This period is the point at which there is a mutual 'testing' process in which attitudes, opinions and opportunities are discussed and debated. In a company with a predominantly formalised product development process, this is the point where the first significant debates arise between traditional and innovative views of design.

Phase 3: integration

The period during which teamwork begins to operate whether the company is project-focused or departmentally driven. For rapid resolution, this period depends on many factors being brought into play, from empathy, the development of a common working language, appreciation of time scales on both sides and a vision of the ultimate design potential of joint activity.

Phase 4: consolidation

Where design personnel become fully integrated and work cohesively and productively with others. In the case of Zellweger, this coincided with intensive structural change as it moved from a traditional management model in which products developed sequentially through independent departments, to a 'matrix' or project-centred model in which projects ran in parallel with company functions organised as 'services' supplied, as and when necessary to each project

The primary factors, which helped to facilitate cultural change through the four phases of the consolidation process, are highlighted in Table 4.

The next section discusses the issue of IT-related design in more depth.

Table 4 The consolidation context

Methods of Encouraging Rapid Cultural Change in Order to Incorporate Strategic in-Company Design Strengths
• Plan a staged, two-way introduction of design into the company • Identify early, small and achievable design goals in order to demonstrate design potential • Adopt a flexible and broad approach to the application of design across the company e.g., examine the potential of a corporate identify programme • Raise the general awareness of design beyond the company e.g., via seminars, small exhibitions, newsletters etc • Work for the design commitment of senior staff and key non-senior staff • Support design decision-making with sound and communicable research where possible • Explore IT-related design opportunities within the company in order to extend design thinking

INFORMATION TECHNOLOGY AS A CATALYST FOR CULTURAL CHANGE

"The 1980s was a decade of high investment in powerful IT-based technologies for automating and integrating different functions, but despite high levels of adoption, contributions to productivity and performance improvements were disappointing. It became apparent that success depended on securing commitment through participative design, and on extensive problem-solving activity around reconfiguring both technology and the organisational context." (Tidd, *et al.,* 1997).

It is often argued that information technology also requires and creates cultural change within companies, frequently of a high magnitude. Less frequently heard is the case that IT can act as a catalytic force for driving disciplines such as design across traditional company boundaries. In the case of the Zellweger TCS, this became a significant factor. At the start of the programme, the Company had a relatively sophisticated IT profile for the period, focused primarily on the engineering-lead areas of technical drawing, information databasing, and project planning; with a relatively unsophisticated three dimensional modelling capability and no multimedia or advanced graphics systems. Whilst it cannot be said that the TCS programme was responsible for the subsequent development of these areas, it is probable that defining the industrial design resource requirements had a substantial impact on IT policy and provided a stimulus for the acquisition, experimental development and application of related resources. It is argued that this created a useful climate within the Company whereby core design data was gradually employed within other areas including: training, promotion and operational testing. This in turn helped create design awareness and improved links with the TCS design team throughout the Company. The potential for this type of IT-stimulated cultural change is summarised in Table 5. This significance of IT as a primary agent of innovation was commented on by Rothwell who appraised different generations of innovation model and concluded with generation five which reflected systems integration and extensive networking, flexible and customised response, engendering continuous innovation (Rothwell, 1992).

Table 5 IT as a Design Facilitator

Creating new design opportunities through:
• Information networks and modes of distribution, increasing design awareness and information flow, independent of traditional company boundaries • New forms of product simulation and visualisation, stimulating ideas, debate & forward thinking • Testing and evaluating 'virtual' products • Increased speed of communication and response – access to all, at any level which supports individual and team involvement • New methods of performing traditional activities, requiring new design skills and allowing design personnel to work with new team partners • IT as a mediator – offering additional credibility to design within the company and a philosophical bridge linking unrelated disciplines

THE DIVERSIFICATION OF INDUSTRIAL DESIGN SKILLS

One of the primary considerations when comparing the effectiveness of in-house, as opposed to external design consultancy, is whether the size and design requirements of the company are sufficient to justify a permanent in-house resource. In the case of Zellweger, the size and range of design demand appeared at the outset to have achieved the necessary critical mass to justify an in-company industrial design facility. However, an unpredicted downturn in the UK economy at the start of the project precipitated a reappraisal. A strategy of diversifying the exploitation of design resources within the Company became a necessity, when the external economic factors of recession began to reduce demand for new product development (NPD) and its associated design requirements. This factor is commented on by Senker who suggested that a very significant factor affecting the benefits which companies derived from their involvement in TCS programmes was the external economic environment. Cash-flow problems or a decline in demand left some companies in a position where interest in TCS programmes or their implementation had become marginal (Senker *et al.,* 1993).

Throughout this difficult economic period, design was initiated in several areas (see Table 6) and a small number of speculative product design projects were also carried out. Temporarily channelling some resources away from NPD, this strategy maintained the momentum of the TCS programme and ensured that it continued to make a positive contribution to the Company.

In retrospect this approach exhibited wider significance beyond external economic factors. It may be the case that manufacturing SMEs which wish to develop an in-house industrial design resource, even within a favourable economic climate and reasonable company performance, cannot guarantee a consistent throughput of new or redeveloped products to sustain such a resource if it is restricted solely to product design. A more divergent view of the design activity may secure the long-term viability of design resources and broaden the design culture within the company. This factor has implications for the recruitment of design Associates, and emphasises the desirability of transferable design skills.

A "SPECULATIVE" PROJECT – A DISPOSABLE DETECTOR

At an early stage in the first Associate programme in discussions with R&D staff, it became evident that there would be possibilities for scaling down the size of some product technologies, sufficiently to facilitate the development of new kinds of product. It was envisaged for example that gaseous detection products could be worn on clothing, as opposed to the relatively bulky existing hand-held products. At this point the company was unsure whether to explore new markets in this area.

Due to the external economic factors and a temporary slow-down in the demand for NPD, it was decided that some rapid, speculative design work should be carried out by the Associate to stimulate discussion and possible future implementation. Design work, followed by presentation models; the 'reality' of a convincing model complete with controls, packaging and corporate logo impressed senior management and played a significant part in stimulating the company to develop and produce a successful product at a later date within a more favourable economic climate. Although general fears were expressed by staff, that such a strategy could possibly run ahead of the abilities of the Company to resolve the related technologies to make the product viable, there was little doubt that such informed and visible speculation could be a useful catalyst for product innovation.

Table 6 A design service throughout the company

Potential Areas for the Diffusion of Design Services:
• **Core industrial design** – Product design, production methods, product graphics, ergonomics, design research
• **Interior design** – office and exhibition
• **Graphic design** – corporate identity, promotional material, web pages, sales literature, training & maintenance manuals, in-house communications, packaging
• **CAD/CAM/CIM** – interface design, rapid prototyping, computer modelling & testing

Table 7 Speculative NPD and design

The effects of a speculative NPD through design visualisation can be to:
• Promote focused NPD discussion across the company
• Act as an opinion test for internal company use or for external user/market/retail research
• Allow designs to be created which respond at an early stage to newly developing product technologies
• Assist with the diversification process by visualising original, state-of-the-art products not previously taken seriously by senior management

It is worth noting that the introduction of sophisticated, three-dimensional CAD modelling during the project considerably increased the potential ease with which speculative projects could be developed and tested.

THE APPLICATION OF INDUSTRIAL DESIGN METHODS TO CAPITAL GOODS

A significant challenge within the programme was the necessity to explore the application of industrial design methods within a company whose products were highly regulated, technologically complex, specialised items of capital equipment. A markedly different design context in comparison with the consumer durables sector where designer labels and styling often sell the product and are therefore often considered to be the primary application of industrial design. In contrast, a number of the Zellweger products including fixed-point detectors, required no human operators once installed, others consisted of standard control racking systems where interface design consisted of rearranging the various control components within the pre-designed systems. Overall, most, if not all, the existing company products exhibited a basic, no-nonsense aesthetic that made few concessions to visual style, product fashion or the vagaries of consumer preference.

This was to a degree the inevitable result of the nature and priorities of the market, which was engineering-based and directed primarily at operational safety. During early discussions it became clear that the Company saw their product 'consumers' as being primarily the specifiers of the products, rather than the end-operators, maintenance engineers or installers. The Associates argued that whilst these types of 'user' were important, it could also be assumed that they would, in many cases, have an understanding of end-user requirements and might well respond to design characteristics which were based on their first-hand requirements. The point was also made that the design of the products had to communicate certain values to the specifier and that it should not be assumed that such a person was any less susceptible to the reassurance of well resolved visual qualities than any other consumer.

In spite of these and other positive reasons expressed for the application of industrial design skills within the Company, there remained understandable doubt about their relevance to such 'functional' products. There was a view that any 'styling' introduced by the designer would be a superficial and rather spurious overlay to the product at the expense of some of more rational, technology-driven detailing already in existence and proven. Previous industrial design consultancy was often cited in support of this argument, and in particular a specific hand-held detector that displayed a particular and distinctive curved profile. It was argued that the production cost of this essentially aesthetic refinement outweighed any real gain in increased sales. There were fears that a superficial 'appliqué' would detract from the essentially functionalist attributes of the products that denoted durability, safety and reliability.

It took a lengthy period of persuasion and the demonstration of innovative design based on user research to change this climate. As a general issue, it raised a potential criticism of industrial design: that it rather neglects the area of capital goods in favour of the more fashion-oriented field of consumer products. This view is reflected in design education, when projects are less frequently directed at the capital goods sector. Thus some design graduates may have little or no experience of the sector, which appears to demand its own particular variants on design insights.

Ultimately, it was the view of the TCS design team that a more in-depth approach to design for the sector was required to establish workable methods of convincingly integrating industrial design skills within the existing engineering and marketing skills of the company. This view is supported by a recent study by the Design Innovation Group (DIG) which examined the role of design and innovation in product competition across real-world companies. In their view commercially successful product development projects, and certainly the more technically complex ones, require a broad, multi-dimensional approach to design with a focus on product performance, features, build quality and, where relevant, technical or design innovation. Loss-making projects tend to involve a narrow, often styling-orientated, approach to design with more attention paid to the product range and cost reduction than to performance, quality and innovation (Roy and Riedel, 1996).

Once again this reflects the need for cultural change, as opposed to simple resource allocation, since a 'broad multi-dimensional approach to design' cannot be achieved by any other means. The factors that may contribute to such an approach within the context of the programme are identified in Table 8.

Table 8 Designing for the capital goods sector

Design abilities which may contribute to design-led cultural change within a capital goods company:

- An understanding of, and ability to design within the companies: structure, markets, competition, technological capacity, resources, production methods and goals.
- Flexible organisational, interpersonal occupational and communications skills.
- Ability to produce design work appropriate to complex, technology-based capital products for professional users.

The final criterion listed in Table 8 is perhaps the most elusive and for which there are no simple answers. Conventional wisdom has it that such products are the preserves of engineering design. The implication is that apparently 'functional' products are designed rather like military equipment with a dominant interest in utilitarian and ergonomic efficiencies, immune from the vagaries of style, symbolism and status more commonly associated with the wider consumer market. However this ignores the expectation that all products have to look what they are in order to be readily identifiable, and to match expectation with the realities of use thus as Baxter suggests in connection with product semantics: "products which move fast should look sleek and streamlined. Products that are durable and hard wearing should look robust and rugged. Products which are fun should look bright and happy whereas products used for serious work should look sombre and efficient" (Baxter, 1996).

The visible attributes for a product do not exist *per se*, a robust piece of industrial equipment for example will not automatically look the part, although substantial clues may be discerned; a product for example reinforced with heavy materials, thick sections and strong fastenings, will clearly signal physical durability. Conversely, it may not be physically necessary or cost effective to design in such "real" attributes, but there may still be a need to reassure the buyer that the product performs to the appropriate reliability level required. In this case, the external form may have to suggest many of the values that the product embodies but are not discernible, for as Oakland suggests "Designing customer satisfaction into products and services contributes greatly to competitive success" (Oakland, 1993).

Clearly this is not an argument for adding spurious visual attributes, particularly in a product sector such as Zellweger's, where safety considerations are paramount and an informed user group might well see through the fake and assume the worst. It might also be considered distinctly unethical and misleading to create product identities that convey safety, robustness and durability if the reality falls short. What is required is a product with the desired optimum performance with a visual identity which matches this, and reinforces appropriate perceptions by the user. Unfortunately, because the design of capital equipment is often perceived as both unfashionable in design circles and without aesthetic requirements, there has been comparatively little investigation or writing which addresses how semantic values in relation to capital equipment can be manipulated successfully by the designer.

Likewise the symbolic aspect of capital equipment, that is its impact on the self-image of the user, is equally neglected. Interestingly, in a related case study conducted by the author into the design of a digital micrometer, preliminary market research indicated that the new product should be aimed at quality control inspectors who commanded relatively high status on the shop floor. The device was ultimately designed to convey a combination of precision and prestige commensurate with the perceived professional role. It might well be argued that there is a greater potential for industrial design in all areas of capital goods manufacture to provide products, both revised and new, with the desired optimum performance in a highly regulated sector; and with a visual identity that matches this and reinforces appropriate perceptions by the user.

IN CONCLUSION

Whilst not without problems, the TCS programme was considered ultimately successful, receiving an alpha rating from the TCS directorate for "scientific / technological merit" and delivering design in both product and organisational terms. The second TCS Associate was successfully employed within the company which has undergone extensive change and is now highly competitive and successful, with a wider portfolio of products and markets world-wide, within which design is making a significant team contribution.

With hindsight, the principal conclusion has been that successful adaptation to, and exploitation of in-house design resources, depended on a combination of attitudinal, cultural and structural adjustments within the company, in parallel with a broadening of the traditional industrial design role by the incoming design practitioners.

It has been demonstrated that the TCS model for engendering change within companies can equally apply to the introduction and integration of industrial design. In this context several key contributors to change have been identified and include: a recognition of the two-way (design / company) processes that successful integration requires; the significance of IT as an agent for cultural change; the need for transferable design skills in the SME environment; with the requirement for increased industrial design subject development to include more knowledge about the design identity of capital goods.

Whilst a single case study may only partially contribute to a generic hypothesis, the issues raised may be relevant to other contexts in which industrial design practice is introduced at a fundamental level within technology-based companies. It is perhaps timely that there is increasing research interest in the lessons to be learned from design-led TCS programmes, in particular current work carried out by the Design Research Centre at Brunel University. In terms of future research, although there has been extensive analysis of the innovation process (see *Journal of Product Innovation and Management*), and some work has been undertaken on the management and practices of companies for "Successful Product Development" (Cooper, 1993); experience in the early stages of the project suggested that there was a comparative lack of knowledge about the activity and organisation of in-house industrial design and a corresponding lack of educational representation. It is hoped that this will be addressed to ensure that further design integration programmes can be undertaken more confidently, with continuous improvement to achieve excellence (Kerzner, 1992; Ulrich and Eppinger, 1995; Wheelwright and Clarke, 1992; Johne and Snelson, 1990; Peter and Waterman, 1982).

REFERENCES

Baxter, M., 1995, *Product Design - Practical Methods for the Systematic Development of New Products.* (London: Chapman and Hall), p.218

Cooper, R. G., 1993, *Winning at New Products - accelerating the process from idea to launch.* 2nd ed., (Addison-Wesley Publishing Company), p.108.

Inns, T.G. and Hands, D, 1988, *Design In-House: 12 case studies describing how the TCS (Teaching Company Scheme) has improved in-house design capability,* (Brunel University: Design Research Centre Publications).

Johne, A. and Snelson, P., 1990, *Successful Product Development - lessons from American and British firms* (Oxford: Basil Blackwell,), p. 147 and 172.

Kay, J., 1993, *Foundations of Corporate Success: How Business Strategies Add Value* (Oxford: University Press, Oxford), p. 416

Kerzner, H., 1992, *Project Management: A Systems Approach to Planning, Scheduling, and Controlling,* 4th ed., (New York Van Nostrand Reinhold).

Oakland, J. S., 1993, *Total Quality Management - the route to improving performance.* 2nd ed., (New Jersey, Nicholes Publishing), p. 43.

Peters, T. J. and Waterman R. H., 1982, *In Search of Excellence: Lessons from America's best-run companies,.* (New York, Harper & Row).

Rothwell, R., 1992, Successful industrial innovation: critical success factors for the 1990s'In *R&D Management,* **22,** (3) pp. 221 – 223.

Roy, R. and Riedel, L., 1996, *The Role of Design and Innovation in Product Competition,* (The Open University), p. 33.

Senker, J. Senker, P. and Hall, A., 1993, Teaching Company Performance and Features of Successful Programmes, *Science Policy Research Unit,* University of Sussex.

Svengren, L., 1998, Industrial Design as a Strategic Resource - A Study of Industrial Design Methods and Approaches for Companies' Strategic Development In *The Design Journal,* Vol. 0, Issue 2, pp. 3-11.

TCS promotional material, 1998, *Teaching Company Directorate,* http://www.tcd.co.uk/.

Tidd, J. Bessant, J. Pavitt, K., 1997, *Managing Innovation - Integrating Technological, Market and Organisational Change,* (Chichester: John Wiley & Sons), p. 322.

Ulrich, K. T. and Eppinger, S. D., 1995, *Product Design and Development.* (New York: McGraw-Hill, Inc.,).

Wheelwright, S. C. and Clarke, K. B., 1992, *Revolutionising Product Development: Quantum Leaps in Speed, Efficiency and Quality.* (New York, The Free Press).

Woolley, M., 1998, Design 'Publications' and the UK University RAE, *Designing Design Research 2 Conference,* http://www.dmu.ac.uk/ln/4dd/drs6.html.

CONTRIBUTORS

Professor Kenneth Agnew, School of Design and Communication, Faculty of Art and Design, University of Ulster, York Street, Belfast BT15 1ED, UK.

Professor Peter Armstrong, University of Sheffield Management School, 9 Mappin Street, Sheffield, S1 4DT, UK.

Karen L. Bull, School of Design Research, Birmingham Institute of Art and Design, University of Central England, Corporation Street, Birmingham B4 7DX, UK.

Professor Rachel Cooper, Research Institute for Design, Manufacture and Marketing, University of Salford, Salford M5 4WT, UK.

Angela Dumas, Judge Institute of Management Studies, University of Cambridge, Trumpington Street, Cambridge CB2 1AG, UK.

Dr. Rosie Erol, Design and Innovation Research Unit, School of Cultural Studies, Sheffield Hallam University, Psalter Lane Campus, Sheffield S11 8UZ, UK.

Dr. Stephen Evans, The CIM Institute, Building 53, Cranfield University, Cranfield, Bedfordshire MK43 OAL, UK.

Andrew Fentem, Judge Institute of Management Studies, University of Cambridge, Trumpington Street, Cambridge CB2 1AG, UK.

Josè Gotzsch, Groupe ESC Grenoble, BP 127, Grenoble Cedex 01, France.

David J. Hands, School of Design Research, Birmingham Institute of Art and Design, University of Central England, Corporation Street, Birmingham B4 7DX, UK.

Dr. Paul Hekkert, Delft University of Technology, Faculty of Design, Engineering and Production, Subfaculty of Design Engineering, Department of Aesthetics, Jaffalaan 9, 2628 BX Delft, The Netherlands.

Eur. Ing. Dr. Bill Hollins, Department of Design, Brunel University, Runnymead Campus, Englefield Green, Egham, Surrey TW20 0JZ, UK.

Avon P. Huxor, Centre for Electronic Arts, Middlesex University, Cat Hill, Barnet, Hertfordshire EN4 8HT, UK.

Dr. Tom G. Inns, Design Research Centre, Department of Design, Brunel University, Runnymead Campus, Englefield Green, Egham, Surrey TW20 0JZ, UK.

Alastair S. Johnson, The CIM Institute, Building 53, Cranfield University, Cranfield, Bedfordshire MK43 OAL, UK.

Sarah A. Jukes, The CIM Institute, Building 53, Cranfield University, Cranfield, Bedfordshire MK43 OAL, UK.

Dr. John Law, School of Product Design, Department of Fashion, Textiles and Three Dimensional Design, Birmingham Institute of Art and Design, University of Central England, Corporation Street, Birmingham B4 7DX, UK.

Dr. Fiona E. Lettice, The CIM Institute, Building 53, Cranfield University, Cranfield, Bedfordshire MK43 OAL, UK.

Dr. Janet MacDonnell, Department of Computer Science, University College London, Gower Street, London, WC1 6BT, UK.

Professor Brian McClelland, School of Design and Communication, Faculty of Art and Design, University of Ulster, York Street, Belfast BT15 1ED, UK.

Ian Montgomery, School of Design and Communication, Faculty of Art and Design, University of Ulster, York Street, Belfast BT15 1ED, UK.

Sophie E. Phillips, Department of Computer Science, University College London, Gower Street, London, WC1 6BT, UK.

Dr. Stephen E. Potter, Design Innovation Group, Faculty of Technology, The Open University, Milton Keynes, MK7 6AA.

Professor Mike Press, School of Cultural Studies, Sheffield Hallam University, Psalter Lane Campus, Sheffield S11 8UZ, UK.

Dr. Johann C. K. H. Riedel, Design Innovation Group, Faculty of Technology, The Open University, Milton Keynes, MK7 6AA, UK. (Now at the Department of Engineering and Operations Management, University of Nottingham).

Dr. Paul A. Rodgers, Engineering Design Centre, Department of Engineering, University of Cambridge, Trumpington Street, Cambridge CB2 1PZ, UK.

Dr. Robin Roy, Design Innovation Group, Faculty of Technology, The Open University, Milton Keynes, MK7 6AA, UK.

Professor Steven A. R. Scrivener, Design Research Centre, Colour and Imaging Institute, University of Derby, Britannia Mill, Mackworth Road, Derby DE22 3BC, UK.

Dr. John. P. Shackleton, Department of Design and Architecture, Faculty of Engineering, Chiba University, 1-33 Yayoi-cho, Inage-Ku, Chiba-shi 263-0022, Japan.

Dr. Heleen Snoek, Delft University of Technology, Faculty of Design, Engineering and Production, Subfaculty of Design Engineering, Department of Aesthetics, Jaffalaan 9, 2628 BX Delft, The Netherlands.

Professor Sugiyama, Department of Design and Architecture, Faculty of Engineering, Chiba University, 1-33 Yayoi-cho, Inage-Ku, Chiba-shi 263-0022, Japan.

Dr. Anne Tomes, Design and Innovation Research Unit, School of Cultural Studies, Sheffield Hallam University, Psalter Lane Campus, Sheffield S11 8UZ, UK.

Alan Topalian, Alto Design Management, 72 Longton Grove, London SE26 6QH, UK.

Dr. Nick Udall, Faculty of Education, Roehampton Institute London, Froebel College, Roehampton Lane, London SW15 5PJ, UK.

Andrée Woodcock, Design Research Centre, Colour and Imaging Institute, University of Derby, Britannia Mill, Mackworth Road, Derby DE22 3BC, UK.

Professor Martin Woolley, Department of Design Studies, Goldsmiths College, University of London, New Cross, London SE14 6NW, UK

INDEX